ANIMAL
PARENTS

**Recent Titles in
Greenwood Guides to the Animal World**

Flightless Birds
Clive Roots

Nocturnal Animals
Clive Roots

Hibernation
Clive Roots

Animal Parents
Clive Roots

ANIMAL PARENTS

■ Clive Roots

Greenwood Guides to the Animal World

GREENWOOD PRESS
Westport, Connecticut · London

Library of Congress Cataloging-in-Publication Data

Roots, Clive, 1935–
Animal parents / Clive Roots.
    p. cm. — (Greenwood guides to the animal world, ISSN 1559–5617)
    Includes bibliographical references and index.
ISBN 978–0–313–33986–8 (alk. paper)
1. Parental behavior in animals. I. Title.
QL762.R66 2007
591.56'3—dc22   2007008803

British Library Cataloguing in Publication Data is available.

Library of Congress Catalog Card Number: 2007008803
ISBN-13: 978–0–313–33986–8
ISBN-10: 0–313–33986–4
ISSN: 1559–5617

First published in 2007

Greenwood Press, 88 Post Road West, Westport, CT 06881
An imprint of Greenwood Publishing Group, Inc.
www.greenwood.com

Printed in the United States of America

The paper used in this book complies with the
Permanent Paper Standard issued by the National
Information Standards Organization (Z39.48–1984).

10 9 8 7 6 5 4 3 2 1

For Jean
For the many years of love, companionship,
and shared concern for animals,
from London Zoo to Vancouver Island

Contents

Preface

Animals must reproduce. Survival and evolution depend upon certain members of every population creating the next generation. They become parents, and the care that many provide is quite obvious. It is clearly shown by the bald eagle bringing fish to its nest, by the mother polar bear nursing her cubs in a snow den, and by the chimpanzee clutching a baby to her chest. Animal parents feed, shelter and protect their young, and may even kill other babies to improve the chances of perpetuating their own genes.

Parental care also includes building nests or digging dens in readiness for the offspring, as well as birds incubating their eggs. It is shown by animals that make special provision to incubate their eggs, like the giant leatherback turtle that swims several thousand miles to a particular beach to lay its eggs, and the iguanas and birds that bury their eggs in volcanically heated soil. Some fish incubate their eggs in their mouths, frogs make foam nests to keep their eggs moist, and pythons coil around their eggs to warm them. Care is provided by animal mothers, fathers, and sometimes other members of the family and even relatives.

These care-providers are the exception, however, for parenthood is not always synonymous with care, and many animals are not good parents. Three-quarters of all fish, most amphibians and many reptiles provide no care at all for their offspring. Some birds are parasites and let others hatch and raise their young, and some chicks are so independent that they raise themselves. Only in the mammals, whose babies must have milk, is maternal care mandatory.

From the garter snakes that scatter at birth, to the mother elephant's many years of tender loving care, animal parents show a wide range of concern for their young. This book is about those parents, mostly the ones who care.

Introduction

Many vertebrate animals are good parents and the responsibilities of parenthood take priority over all other aspects of their behavior. Parental care becomes the dominant force in their lives. They make nests and dens in preparation for the arrival of their young or for the incubation of their eggs, and they feed their offspring and protect them from predators and the elements. The actual type of care they provide is closely connected to the mode of reproduction practiced by their species, especially whether they lay eggs or give birth to live young. The egg-layers include most fish, amphibians and reptiles, all birds, and a few mammals.

Laying eggs provides varying opportunities for parental care. Eggs have their own internal supply of energy and hatch after they are incubated, if they are fertile. In the ectotherms (the cold-blooded fish, amphibians and reptiles), the environment provides the conditions for the eggs' embryonic development; incubation by the mother is therefore unnecessary, and impossible anyway as she cannot provide warmth. Parental care is then mainly concerned with protecting the eggs until they hatch, and then, in a few species, of caring for the young. In the warm-blooded birds and the monotreme mammals (the echidnas and platypus), parental care of the eggs involves their incubation through warmth, and most species then provide food for the young. These animals, and the reptiles, lay shelled, amniotic eggs, in which a tough, membranous sac encloses the embryo so they can be laid on land, unlike the desiccation-prone eggs of the water-dependent fish and amphibians. They can lay them and leave them, and many reptiles and a few birds do just that; but this does not mean they are poor parents, for they have provided for them beforehand.

In the animals that give birth to live young, one of four different methods is involved; two each in the ectotherms and endotherms. In the ectotherms, both methods involve the development of the egg within the mother's body. Ovoviviparity is the name given to the mode in which the embryos are fully developed when the eggs are laid and immediately break free of their membranes. This is considered a form of live-birth and occurs in several fish, a few amphibians, and in many snakes and lizards. However, the embryos are nourished entirely by the egg yolk and do not receive any nutrients from their mother. The other form of live-bearing in the ectotherms is called viviparity and occurs in some fish, amphibians, and reptiles. Here, the mother provides nourishment to the embryos while the eggs are still inside her body, similar to the development of higher mammals but without a placenta, and then gives birth to live young. In some species the eggs hatch inside the mother and the larvae are nourished by her, or by eating their siblings. The other forms of live-birth involve only the mammals. First, the marsupials, which lack a placenta, although a very temporary connection supplies some nourishment to the embryo; but the babies are always "premature" and must then be nursed in the mother's pouch. Finally, the placental mammals, in which the young are nourished in the uterus via the placenta and are well-developed at birth.

Giving birth to live young provides fewer opportunities for care. Ectotherm young tend to scatter at birth. In the mammals, however, parental care is mandatory as the young must be nursed, although their requirement for warmth depends upon their degree of development—whether they are altricial (naked and helpless) or precocial (furred and mobile). When the young are born or the eggs hatch, parents must protect, shelter, and feed their offspring. Parental care is therefore an aspect of reproduction, involving behavior that occurs before and after birth. The breeding strategies that occurred earlier, such as display, pair formation, and then mating, while obviously part of reproduction, are not aspects of parental care. Similarly, the internal functions between the two phases—such as the fertilization of the egg and the development of the mammalian embryo within the mother—are not considered parental care. They are mentioned briefly only when they are relevant to the subject of this book—the varying degrees of care provided by animal parents that improve the survival chances of their offspring.

■ THE PHASES OF CARE

Parental care is generally assumed to commence when eggs hatch or young are born. But for many animal parents it actually begins long before that event, with the preparation of the nest or den and the incubation of the eggs. Consequently, there are two distinct phases of care, pre-birth and post-birth.

Care before Birth[1]

Before the arrival of their young prospective parents must prepare dens or nests. Many must lay eggs, and in most cases arrange for their incubation, either by themselves, by others or by the environment. Pre-birth care occurs in all the classes of vertebrates—the fish,[2] amphibians, reptiles, birds and mammals. Examples of care before birth include the burrows dug by penguins, aardvarks, and warthogs; the nest mounds built by crocodiles and megapodes; the cavities drilled by woodpeckers; the huge nests of branches made by bald eagles; and the dens made by northern bears in preparation for hibernation and raising their cubs. This protective behavior is concerned mainly with providing the environment in which to receive the offspring. It may occur before mating, as with the fish and amphibians that build a nest and then attract a mate. In birds, nest building generally coincides with mating, as the eggs must have several days to develop after they are fertilized. Or it may occur long after mating; placental mammals give birth many weeks or months after they have mated, and then close to parturition time must prepare their den or nest.

Pre-birth care has a more significant role in the egg-laying animals. Producing eggs does not constitute parental care, only an investment of energy by the mother; but caring for the eggs is a major aspect of pre-birth care in many species. It includes their protection and incubation by the birds, and their hydration by certain frogs. Other frogs carry the eggs on their backs, several fish "incubate" them in their mouths, and the seahorses carry them in stomach pouches. Pythons coil around their eggs to incubate them, and crocodiles make nests for their eggs and then guard them. But of all the ectothermic vertebrates, only the pythons provide warmth and thus incubate them in the accepted sense.

Although nest building is the obvious form of parental care prior to egg-laying and incubation, many birds do not even make a nest for their eggs. The male emperor penguin incubates his mate's single egg on his feet and boobies wrap their large webbed feet around their eggs to warm them. Many seabirds lay their eggs on a bare rock ledge or directly onto a sandy beach, and fairy terns balance their single egg on a branch. Other animals produce eggs with incorporated survival aids, or they find a site that will improve the chances of successful incubation, or both. They therefore provide good care prior to egg-laying, but not afterward. Some megapodes, chicken-like birds that rely on the environment to incubate their eggs, not only find suitable volcanically-heated soil, but lay large-yolked eggs that have a long incubation period so that when their chicks hatch, they are independent, well feathered, and able to fly. As a group, the reptiles actually provide considerate care for their eggs. Some snakes, crocodilians, and lizards guard their eggs, but with few exceptions, all the other egg-laying reptiles bury their eggs to protect them from predators and the elements, in a place where the conditions are most suitable for their incubation. Some go to great lengths to find the right place—a sun-warmed beach, a pile of decomposing vegetation, soil heated by volcanic activity, or an animal burrow—all of which constitute pre-birth care.

Pre-birth care is absent in many mammals, however, although they are very caring after birth. Female primates, except for the nest-building prosimians, simply give birth and their babies cling to them. Most ungulates or hoofed mammals make no preparations for the arrival of their lambs and calves, other than perhaps isolating themselves from the herd for the actual birth. In most species the young are mobile and keep up with their mother within a few hours of birth. The pinnipeds (sea lions and their relatives) give birth on a beach without any preparation, other than in some species migrating a very long distance to reach the beach where they were born. The major differences in pre-birth care occur between the endotherms and the ectotherms. The monotremes and all birds except the megapodes provide body-warmth for their eggs during incubation. The ectotherms cannot do this, as they cannot generate heat,[3] so their caring is for protection only, not incubation in the normal sense. Also, the incubating bird may be fed by its mate, whereas such behavior is unknown in cold-blooded animals.

Care after Birth

The duties of parents are basically twofold—feeding and protection. Generally the amount of care increases as the complexity, size, and lifespan of the animal increases. All the classes of vertebrates provide examples of parental care after birth, with the major differences again being between the endotherms, or warm-blooded birds and mammals, most of which provide warmth for their young; and the ectotherms, or cold-blooded fish, amphibians, and reptiles, that cannot even generate their own body warmth. In the ectotherms, only the discus fish and some amphibians feed their young, whereas most birds are fed by their parents or are shown where to find food, and all baby mammals are initially dependent on their mothers or a foster parent for milk. Some endotherms also bring food to their partners while they are caring for the offspring. Post-birth care varies considerably in length, from just a few days for the mouth-brooding cichlid fish and the hooded seal, to eight months for the wandering albatross, and several years of care by African elephant and lowland gorilla mothers.

Parental contributions to the survival of their offspring take three forms—female care, male care, or biparental care. The latter is most common in birds, in which 90 percent are monogamous (whereas only just over 3 percent of mammals are) and both parents share the incubation and brooding. Providing care is a question of costs and benefits. In the vertebrates, the great variety of types have resulted from evolutionary pressures relating to behavior beneficial to the offspring versus the costs to the parents. The benefits to the offspring are protection, food, and in the endotherms the provision of warmth, all of which improve their chances of survival. These benefits aid the individual, its population and its species. The parents benefit from passing on their genes to the next generation.

The major cost to the parents is the investment of energy they must make, because providing care for the young is a costly business. An animal's body needs energy to function; and to produce young and then raise them successfully to independence requires extra energy. Mothers can mobilize some of their offspring's needs from within their own bodies—calcium, for example, to create eggshells or their embryo's bones—but this must eventually be replaced, and in mammals the production costs of the developing fetus are very high. There are also more direct costs to the physical health of some species, such as the exposure risk to predation of the nestlings and their parents in an open nest, and pythons coiled in full view around their clutch of eggs. In addition they may eat improperly and lose condition. Male emus may not leave their nest, even to drink, for the whole 56-day incubation period and can lose up to 22 pounds (10 kg), almost one-fifth of their bodyweight. Some incubating animals, like the cichlids with a mouth full of eggs or fry (hatchling fish), simply cannot eat. Finally, investing in a family results in missed opportunities. While she is raising her young, the mother misses the opportunity to produce more young, and in many species the male loses the chance to mate with

Giraffes *Giraffe calves are very vulnerable to lions and hyenas.*
They stay in a creche with other calves in the bush initially for
several months, then join their mothers and young bulls in open
country where predators cannot creep up on them unseen, but
food trees are sparse. The larger bulls, at less risk from predators,
can stay in the bush where food is more plentiful.
Photo: Vladimir Kondrachov, Shutterstock.com

other females and spread his genes—for the most critical resource for males is
access to females, whereas for females it is access to food.

The parental investment or energy costs of reproduction vary according to the
biology of the species. The major investment in mammal reproduction begins at
conception and lies in the development of the fetus. After birth the investment costs
for mammals are reduced to supplying milk and, for some species, the need to keep
their babies warm. In birds, at least in those with altricial young, the costs are
greater after the eggs have hatched than the energy expended to lay and incubate

them, with the nestling's demand for food taking up much of the parent's daily activity. These demands drain a bird's reserves. In addition to the constant pressure from their hungry chicks, the birds must also protect them and contend with the rigors of daily life, such as escaping their enemies and defending their territory, during a time when they are probably not eating properly themselves. All these factors reduce the birds' own chances of survival.

The hormone prolactin is connected to parental care. It is most well known for its role in producing milk protein, and is essential for lactation in mammals. While it plays a very important role in maternal care, it is also involved in paternal care, which occurs exclusively in some fish, amphibians, and birds. It is closely related to the growth hormone and is produced in the anterior pituitary gland, but is also secreted by other cells in the body. In mammals, prolactin targets the mammary glands, stimulating their development and then lactogenesis—the production of milk after birth. In birds, prolactin levels rise during egg-laying, peak during incubation and then decline when the eggs hatch and other stimuli—visual and vocal—prompt the feeding response. In male birds, prolactin increases during incubation and has been called the paternity hormone, and higher levels of the hormone even occur in non-breeding birds that help at the nest. Like the development of the mammal's mammary glands, which are milk-secreting adaptations of sweat glands, production of the pigeon's crop "milk" is controlled by prolactin. High prolactin levels have also been found in male fish—in the North American bluegill (*Lepomis macrochirus*) and the three-spined stickleback (*Gasterostreus aculeatus*), both of which show exclusive paternal care of the young.

In most cases parental care of the eggs or offspring involves "external" care. Fish, amphibians and reptiles guard their eggs, and in a few cases their young. Birds incubate their eggs and raise the chicks, and carnivores care for their cubs or pups in a den. But there are also many examples in the animal world of "internal" care. It occurs in the amphibians, whose eggs are incubated while embedded in the skin on their backs; in the male sea horses that incubate their mate's eggs in a stomach pouch; and the most famous examples, the marsupials, who raise their tiny newborn young unseen in the protection of the mother's pouch.

Parental care occurs in varying degrees in all the classes of vertebrates. It is surprisingly well developed in fish, with about one-quarter of all species providing some form of care. This percentage drops in the amphibians, and it reaches its lowest point in the reptiles, where very few species provide any care for their young but many care for their eggs. In theory, the differences in the methods of parental care and therefore of evolution result from the varying benefits that the offspring receive versus the cost to their parents of providing that care. Although the basic functions of care-giving have certain similarities in many animals, parental care in the vertebrates differs considerably between the classes (such as between birds and mammals), between the orders (carnivores and primates), between families (lemurs and anthropoids), and even within families (chimpanzees and orangutans).

The other major aspect of parental care, almost as important as feeding and protection in some species of birds and mammals, is teaching their young how to survive. This is especially important for the carnivores that are not large enough to hunt successfully when they are weaned, and becoming accomplished hunters requires both training and growth. Polar bear and tiger cubs stay with their mothers for two years and are taught how to hunt. Spotted hyena pups are weaned when one year old, but more months pass before they become successful hunters.

■ SPECIFIC CARE

Ectotherms

About one-quarter of all fish show some form of parental care, and it may be the duty of the male, the female, or both. With one exception—the discus fish—fish do not provide food for their young, so one parent is normally sufficient to care for the eggs and then guard the fry. Male care includes nest-making, attracting females to lay their eggs, and then guarding, aerating and cleaning the eggs. In several species it involves skin brooding, pouch brooding, or mouth brooding. Female care also includes mouth brooding the eggs and larvae. Most bony fish (*Osteichthyes*) are egg-layers, and fertilization is mainly external. The young, called fry or larvae, are neither fully-formed nor free-swimming when they hatch, and still depend on their attached egg sac for nourishment. In these species, the male is the primary caregiver.

In the amphibians, parental care is rare in the tailless frogs and toads (*Anura*), with only about 10 percent of the almost 5,300 species caring in any way for their eggs or offspring, and in some species for both. It is more common in the tailed salamanders (*Caudata*), in which 18 percent of the 555 species care for their eggs. The care provided is mostly external in nature and includes the creation of foam nests by many frogs to protect their eggs, physically guarding them and moistening them with urine to prevent their desiccation. Internal care occurs in several species when the eggs become embedded in the parent's skin and develop into froglets there, and some species even carry their tadpoles on their backs. Like the fish, when amphibians provide parental care, one parent can cope, as only a few frogs feed their young—laying eggs for them to eat.

In the reptiles, direct care of the eggs or young occurs only in the crocodilians, pythons, and a few other snakes, the snakelike glass lizards, and the skinks. Crocodilian behavior includes both pre-hatching and post-hatching care—making a nest, guarding it and the eggs, and then carrying and guarding the hatchlings. Both parents are involved, but in the other care-giving reptiles maternal care predominates; in fact, there are few examples of paternal care. Female pythons coil around their eggs and can raise their temperature slightly to incubate them. The glass lizards and the skinks curl around their eggs, and skinks may also protect their hatchlings for a while, mostly from other skinks. The large prehensile-tailed skink is a live-bearer and guards her single big baby. No reptiles provide food for their young.

The evolution of the amniotic egg—in which the skin or shell is impervious, therefore reducing water loss—gave the reptiles a great advantage over the fish and amphibians, allowing them the opportunity to lay their eggs terrestrially. Consequently, many species of tortoises, turtles, lizards, and snakes bury their eggs as they lay them, protecting them and providing the environment needed for their incubation. This kind of care is obviously the female's task.

Endotherms

Parental care by birds is highly developed in most species and involves care of the eggs and the young. The major difference between their young, and the one that most affects the manner of their raising, is their degree of development upon hatching. They are either altricial (helpless) or precocial (well developed). The evolution of their mating systems to a mainly monogamous one has resulted in the dominance of bi-parental care (in 90 percent of all birds), in which both parents share the incubation and chick-raising. Females alone care for their young in only 8 percent of birds; and males are the sole care providers in just 2 percent. Birds were the first vertebrates in which all species lay eggs, and most prepare nests beforehand and raise their young to independence. However, paternal care is just as possible as maternal care by birds once the eggs are laid—unlike the mammals, in which there is no natural alternative to the mother nursing her young, and paternal care is therefore restricted to other duties such as carrying, grooming, and protecting.

Only a few birds do not provide care for their eggs. They are the megapodes that bury their eggs in soil heated by the sun or geothermal activity,[4] the black-headed duck, and numerous other parasitic species. The megapode's chicks and the black-headed ducklings are totally independent when they hatch. Most other birds feed the young they have hatched, whether they laid the eggs or not. Those with altricial chicks bring food to their young in the nest, while those with precocial chicks either offer them food, show them where the food is, or just let them find their own. Some birds have evolved highly specialized food for their young. Flamingoes, pigeons, and penguins all provide the equivalent of milk, a highly nutritious secretion of the crop or esophagus, for their chicks. This behavior evolved separately to meet the specific needs of their species.

In contrast to the ectotherms, birds must keep their young dry; even ducklings must be kept dry when they are out of the water. They must also be kept warm, their body temperature maintained at the fixed level for the species, by balancing heat production and heat loss—the process of thermoregulation. This is achieved by brooding—almost continually in the case of naked and helpless altricial young in the nest; and frequently, especially at night and during bad weather, for down-covered precocial chicks.

Mammals differ from all other animals in many ways, but the major one, at least where parents are concerned, is their production of milk. They are the only vertebrates in which the mother's role in raising the young is essential, unless replaced by foster care or artificial hand-raising. The evolution of parental care in the mammals began with the egg-laying, partially reptilian monotremes or prototherians, then advanced to the marsupials or metatherians, and then to the placentals or eutherians. The baby monotreme hatches from a tiny egg, but is then suckled with milk—combining the reptilian characteristic of egg-laying with mammalian milk production. The short gestation period of the marsupials results in literally an external embryo, whose growth depends on milk for much longer than most placental mammal babies, and the composition and volume of marsupial milk changes during the lactation period to compensate for the stages of growth.

In the placental mammals, the young are more advanced at birth, due to the mother's investment in them during their lengthy gestation. This is followed by a lactation period that may last several weeks, months, a year, or in the case of the African elephant at least three years, but on average is much shorter than the marsupial's nursing period. Long gestation periods are made possible by the presence of a placenta, which transports nutrients from the mother to her offspring and allows waste products to be excreted. The composition of mammalian milk varies considerably according to the lifestyle of the species. It may be very dilute (low in solids) like the milk of the American bison, whose calf stays close and drinks frequently; or have a fat content of over 50 percent, as with the seals who abandon their pups on the ice after nursing them for just a few days.

As milk is strictly a female product, paternal care in mammals is obviously not highly developed and males provide care after birth in only a few monogamous species. The dwarf hamster, gibbons, and the Patagonian mara are examples of these, in which the male is normally present throughout the gestation period and assists in the raising process as a guardian and food provider; the hamster even fusses around while his mate gives birth. This behavior is usually considered a case of the male protecting his genetic involvement in the family, as he can be almost certain the young are his own. Most mammals, however, are polygynous, in which a male mates with several females and can only protect his offspring in social species that live in herds, like the zebra, and not in wide-ranging solitary ones, like the tiger. A few mammals, such as the marmosets and lions, have a cooperative system in which breeding is restricted to one or two females, and several adult males and other females assist in caring for the babies.

The major differences in mammalian parental care occur as a result of the condition and needs of the baby at birth. In this book the mammals are divided into five categories, three of altricial species[5] and two that have precocial babies. The altricial species are the implacentals, the helpless young that are born in nests or dens, and the almost as helpless ones that are carried by their parents. The precocial species are those in which the babies are generally perfect miniatures of their parents, and can follow them soon after birth; plus the young of the pinnipeds, so well developed and receiving such rich food, they can be left on the beach or pack ice for days.

■ WHO CARES?

There is little variation in who provides the care in vertebrates, for it can only be the mother or father or both, and in a few cases other helpers, but females are the primary care-givers. In the mammals it is unquestionably the mother initially, for only she can provide the nourishment the newborn needs for its survival. As to why male care, in addition, is uncommon in mammals hinges on the question of paternity, for the male gains nothing from caring for the young if he is not their father, and it is therefore more advantageous for him to find other females to mate, which is the more typical polygamous arrangement. Then there is the question of association with the offspring. When the eggs are fertilized internally, the male can abandon the female and seek other mates long before the eggs are laid or the

Royal Albatrosses *After sharing the incubation of their single egg for about 11 weeks—the longest of all birds except the kiwi— royal albatross parents then take turns brooding, guarding, and feeding their single chick for its first five weeks. Then they both go to sea to keep up with the chick's demands, returning every two or three days with food. In total they care for their egg and chick for almost 11 months.*
Photo: Silense, Shutterstock.com

young are born, leaving the female totally responsible for birth and raising. It alters the situation somewhat if fertilization is external, as in many fish and amphibians; for the male must be there to fertilize them, and he may then get left with them.

One fact is indisputable, however: parents care for their own offspring, and rarely the young of others. The unnatural cases of human-instigated fostering of domesticated kittens to a lactating dog, or allowing a tigress to raise lion cubs, would not occur in the wild. But intraspecific fostering does occur in some species, such as the female monkey that cares for another's baby, and the walrus that may adopt an orphaned calf. It is also well documented that some birds may be so stimulated by the calls of chicks in another nest that they cannot resist feeding them.

However, the characteristically human behavior of feeling kindly to babies is unnatural in wild animals, and many regard the young of others, and sometimes even their own, as food. There have been rare cases of interspecific "fostering," even in animals that normally have a predator-prey relationship. This happened recently on three occasions in Kenya's Samburu National Park, where a lioness adopted oryx calves, protecting them but allowing them to return to their mother to nurse. This strange association was unsuccessful as the first two calves were eventually eaten by less caring members of the lioness's pride, and the third one was removed by rangers as its mother was too alarmed by the presence of the helper to feed it.

When raising their families, some parents receive help from others, who have been called aunties, helpers at the nest, co-operative breeders, or parent's helpers. These helpers show altruistic behavior known as alloparenting, that benefits others of the same species and boosts the recipient's survival chances and future reproduction, but at some cost to the helper's own breeding potential. It occurs in group-living species, such as the meerkat (*Suricata suricatta*) and the Florida scrub jay (*Aphelocoma coerulescens*), in which females forego the opportunity to breed and help others in the group rear their young. As evolution results from maximizing reproduction, this behavior puzzled zoologists. But it is now believed that this "babysitting" stems from a shortage of resources, resulting in cooperative foraging. This improves breeding success by fewer mothers, but as they share many of their genes with the related helpers, the group benefits anyway. In return, the helpers receive experience in parenting for when they eventually have their own young, at which time they can expect reciprocal help.

Some birds exploit others, often of completely dissimilar species, by forcing them to undertake all their parental care duties. Called brood parasites, they lay their eggs in other birds' nests and let those birds incubate them and raise the chicks. Five families of birds contain these parasites, and this behavior allows them to lay many more eggs than normal, as they are not restricted to the number of eggs they could incubate or chicks they could raise, thus saving energy. The parasitic bird may remove the host bird's eggs or pierce them and kill the embryos, or the parasite's chick may kill the host bird's chicks or push them out of the nest. In a few species, the parasitic chick may grow up alongside the legitimate chicks. A rather surprising aspect of this behavior is the often striking difference in size, and sometimes in color, between the eggs of the parasitic birds and their hosts. Consequently, parasites' eggs are sometimes rejected, but generally the success rate is high, and parasites are considered very successful birds. Another form of parasitism, called "conspecific brood parasitism" occurs in ducks and geese that lay their eggs in the nests of females of the same species, while also probably nesting themselves.

■ YOUNG AND HELPLESS

Parental care involves more than just providing food and shelter. Baby animals must also be protected from the many predators that seek to eat them, but for many species the danger lies much closer to home. The practice of infanticide—the killing and eating of their infants, or those of their conspecifics—occurs in many ground squirrels and other rodents, especially rats, and in primates, lions, and several birds of prey. A rodent mother may cannibalize her own young if she is very stressed and fears for their safety. In some colonial-nesting species, such as gulls, nonbreeding birds kill and eat nestlings, a practice that may stem from overcrowding. Many adult snakes consider smaller snakes food, even their own offspring from eggs laid several weeks earlier. After guarding its eggs during their incubation, the cannibalistic king cobra leaves them just before they hatch—supposedly to avoid the temptation.

Entellus langurs, large long-tailed monkeys of northern India, normally live peacefully. But when a strange male assumes control of a troop—by replacing an existing male through a challenge or by his death—he kills all the troop's babies that were sired by his predecessor. This also happens in lions when a new male takes over a pride and kills all the young cubs unless the lionesses are able to hide them or gang up against him to protect their babies. Polar bear cubs must avoid adult males, even their own father, who regard them as prey. Pups of the harem-forming seal lions and fur seals may be killed by a neighboring "beachmaster" bull merely for straying into his territory. Another lethal hazard for some young birds is the practice of siblicide, which occurs in several raptors, herons and boobies, in which the youngest chick succumbs to its aggressive sibling. The only case of siblicide in mammals occurs in the spotted hyena.

■ TERMINATION TIME

In all animals there comes a time when the bond of dependency between parent and offspring must be terminated, and in most species of birds and mammals there is conflict between the parents and their young over when to do this. As evolution favors strategies that maximize reproduction during an animal's entire breeding life, rather than a single season, care given to a particular brood may be limited. So the parents may need to terminate the care before it is in the best interests of their offspring to leave home. Conversely, while longer care improves the babies' chances of survival and their eventual reproductive success, it may jeopardize the parents' ability to raise a second brood before winter sets in, or allow them to recover their condition before they must migrate or hibernate. But the time arrives when the cost to the mother of continuing to provide care outweighs the benefit to her offspring, and parents must often physically drive their young away or ignore them. It is common, even at the garden bird table, to see a full-fledged nestling pester its parents with the typical begging gestures of shuddering wings and gaping, and then pick up seed itself when it is ignored.

Notes

1. Births are considered to be the hatching of the egg or the live birth of an animal.

2. There are actually three classes of fish. The Chondrichthyes are the cartilaginous sharks, skates and rays; the Agnatha are the hagfish and lampreys, and the Osteichthyes are the bony fish, which includes most of the species. They are referred to collectively here as fish.

3. Except for brooding pythons, which coil around their eggs and can raise their body temperature several degrees through muscle contractions.

4. The mound-building megapodes care for their eggs by regulating the temperature of the mound of soil and vegetation.

5. Although the terms "altricial" and "precocial" correctly apply to the condition of animals at birth, they are also used generally to denote the species that produce these babies.

1 Water Babies

Fish were the first vertebrates to inhabit the earth. Their evolution began about 510 million years ago in the Cambrian Period with creatures whose bones comprised only a skull and a partial covering of armored plates, followed by the first sharks and bony fishes 100 million years later. One could naturally assume that parental care continually improved throughout their evolution up to the present day, but apparently this was not the case. Many transitions between care and no care occurred during that time, although it is generally accepted that egg-laying was the ancestral state and that male care evolved first. All fish need water to survive, either all the time or seasonally in the case of some estivators that lie dormant in the dried mud at the bottom of their pond. With few exceptions, their eggs cannot survive out of water as they are very vulnerable to desiccation, even those of the sharks with their horny, waterproof covering. Caring for their offspring seems unlikely behavior for fish, in view of the nature of their environment and the fact that they are cold-blooded. Yet 25 percent of the almost 24,000 species are caregivers, so the world's first vertebrates actually show more interest in protecting their eggs and young than the amphibians and reptiles that evolved later.

There are three groups or classes of fish. The primitive jawless fish like the parasitic lampreys belong to the class *Agnatha,* and the fish with cartilaginous skeletons such as the sharks and skates belong to *Chondrichthyes.* All the others, most species in fact, are classified *Osteichthyes;* they are known as the bony fish and are further subdivided into the ray-finned fish, such as the sturgeon, catfish, and trout; and the lobe-finned fish, like the coelocanth and the lungfish. The physical care of their eggs or young externally occurs only in members of the bony fish group. They are the subjects of this chapter and will be referred to collectively as fish.

Parental care is an aspect of reproduction, and the first stage of that process reveals the mating systems that determine the role of the individuals and consequently influence the care provided. Monogamy, when one male and one female mate for a long period, is rare in fish and is considered the most primitive mating strategy. Seahorses are monogamous and bond for life—which is also unusual in fish. Polygamy is the most common strategy and takes two distinct forms. When a male mates with several females—such as in the African cichlids, in which the male defends the nest site and attracts females—it is called polygyny. Females lay their eggs in his nest, and he then fertilizes and cares for them. When several males mate with one female, which is very rare behavior in fish and occurs only in species whose males can brood just a few eggs at a time—such as the pipefish—it is called polyandry. Promiscuity, when many males and many females mate indiscriminately, occurs mainly in pelagic species like herring and cod. Lek mating, an unusual variation of breeding behavior in which males congregate to attract females, and which is more common in birds, occurs in a Lake Malawi cichlid (*Copadichromis eucinostomus*). A thousand males of this species congregate and build a large "nest" of sand to attract females. They may mate with several males, laying a few eggs for each partner to care for—in the typical cichlid manner of mouth brooding.

None of the fish that give birth to living young care for their offspring. Only the egg-layers provide parental care, of both their eggs and then the fry (the larvae or hatchlings), although most abandon their eggs as soon as they are laid. The pelagic marine fish lay many undeveloped eggs that are either carried by the current across a wide area, or drop to the ocean floor. Female cod, for example, may produce several million eggs at one time, and are known as "broadcast spawners," their eggs being fertilized externally (the system that most fishes employ); the males releasing their milt (sperm) to coincide with the females laying their eggs. There is no parental care in these fish, and there is heavy predation of both eggs and fry, with a loss rate as high as 99.9 percent of the eggs laid. This is exactly why they lay so many eggs, to compensate for the expected high losses.

Fish that do practice parental care counter predation in another way—by producing fewer eggs and reducing the risk of losses through their care, so the end result is much the same for both kinds. Generally, the fewer eggs laid, the more likely parental care will be involved, and its evolution in fish has resulted in a number of unique methods of guarding their eggs and, in some cases, the fry. Larvae are less vulnerable as they can move, and scatter when threatened, yet egg-laying is the most common reproductive strategy in fish, a practice called oviparity. Fish eggs have a soft, transparent, and permeable shell, and waste products diffuse into the water and respiratory gases are obtained by diffusion. In most species, fertilization occurs externally after the eggs are laid, and they then develop outside the mother's body.

The lack of parental care in fish was likely their ancestral condition, which evolved into male care as they guarded the eggs while guarding their territory. Caring for their eggs is generally not a sound investment for females anyway, as it reduces their fecundity or reproductive potential. As fish continue to grow for all or most of their lives, the energy spent in caring for eggs or fry instead of eating, growing, and being able to lay larger broods of eggs would be a poor investment. When males guard the eggs, they may also attract other females who lay their eggs alongside, allowing these females to invest more energy in their own development, rather than in providing parental care. The parental duties of fish include making nests for the safety of their eggs and young, physically guarding the nests, cleaning the eggs, fanning the eggs to increase oxygen circulation around them, and removing dead eggs. It also involves "brooding" the eggs and fry, either in the mouth, in a brood pouch on the abdomen, or in their skin, guarding the fry in an external nest and retrieving any that scatter. As few young fish are provided with food by their parents and none receive warmth, a single parent is usually sufficient to care for the offspring. When males evolved caring behavior, the females were free to produce more eggs.

Fish protect their eggs in a number of ways. Egg-buriers include species that live in seasonally dry ponds, such as the killifish that lay their eggs in the mud where they remain dormant until the next rains. Some fish lay their eggs in caves where they are difficult to find; others, like the rainbow darter (*Etheostoma caeruleum*) of the eastern United States, bury their eggs in gravel on the stream bottom. The burrowing gobies (*Trypauchenidae*) take parental care to another stage; the female lays her eggs in a burrow, and a male enters to fertilize them. She seals him in for several days, although she does return periodically to open up the burrow—to lay more eggs, not to feed him—and then reseals it and leaves. Eventually just one juvenile leaves the burrow with the male, so it is assumed that the single baby is cannibalistic and grows at the expense of its siblings, or that the hungry male ate them. A more common form of fish parental care is nest-making, which

Angel Fish   *A cichlid from the Amazon Basin, and a very popular aquarium species, parent angel fish clean off a flat rock, lay their eggs there and then guard them until they hatch. Their parental duties also include cleaning the eggs by sucking them into their mouths and then spitting them out.*
Photo: Rui Manuel Telez Gomez, Shutterstock.com

varies from a simple depression in the substrate to a nest of bubbles. The depression, made by sucking up gravel and sand and then blowing it away, is practiced by a number of fish, especially the cichlids. Many gouramis, including the popular aquarium fish the Siamese fighting fish (*Betta splendens*), make floating nests of saliva-coated bubbles for their eggs, which the male aggressively defends after he has fertilized them. Others, such as the spotted betta (*Betta picta*), are mouth brooders. The three-spined stickleback (*Gasterostreus aculeatus*) glues together pieces of algae with its own body secretions, so that it resembles a more typical nest. The females visit the nest to spawn, and the males then protect the eggs from predators, especially other sticklebacks.

Many other fish do not make nests, or even simple depressions, but may guard their eggs wherever they have laid them. These egg-depositors, as they are called, attach their eggs to rocks or aquatic vegetation and then either provide care, like the cichlids, or do not, like most of the bony fish. Although most pelagic fish broadcast their eggs, marine species living along the shoreline or at least close to the shore lay fewer eggs and then protect them. The strange-looking lumpsuckers (*Cyclopteridae*), tiny round fish with an adhesive pad behind their pectoral fins for attaching themselves to rocks in the cold waters of the northern continental shelf, spawn in shallower water near the shore, often in old shells. The males then guard the eggs until they hatch, fanning them with their fins to circulate oxygen. Within the large group of gobies—almost 2,000 species with marine, freshwater, and brackish water representatives—the males of most species guard their eggs until they hatch.

There are also many examples of parental care in freshwater fish. In behavior resembling the creching habits of certain birds, the African catfish (*Clarius gariepinus*), a species commercially bred in many countries for food, gathers the fry of other fish into a school with its own fry, but for purely selfish reasons. It herds the unrelated young to the outside of the school where they are the first to be picked off by predators. Males of the North American bluegill (*Lepomis macrochirus*) nest colonially and females visit the site to spawn. The males then protect them from predators, and fan them to improve oxygen circulation. The round goby (*Neogobius melanostomus*) of the slightly salty Caspian Sea but introduced and established in the Great Lakes, is also a caregiver. The female lays up to 2,000 eggs on a rock, and the male aggressively defends them and the larvae. Male largemouth bass (*Micropterus salmonoides*) a freshwater sunfish, make a nest in the gravel in shallow water in small lakes and slow-running rivers. They defend the nests in which several females may spawn, each providing several thousand eggs. The male fans them to aerate them and guards the fry for several days after they hatch.

The most advanced care behavior in fish occurs in the brooders. Most hold their eggs and then the fry in their mouths to protect them, but some care for them in skin pouches and even in their gills. The cichlids are the most highly developed of these caregivers, providing sanctuary for their eggs and fry in their mouths, and are consequently called mouth brooders or oral brooders. They show an amazing range of mating systems, feeding behavior, and parental care, and the group is considered one of the finest examples of vertebrate evolution, plus one of the most popular groups of aquarium fish. There are hundreds of species of cichlids, and classifying them is very difficult because there are so many examples of convergent evolution, where similar traits have developed in unrelated fish that adopted similar lifestyles. The many species inhabiting the isolated lakes of East Africa show the most diverse adaptive radiation of all fish, with many changes in appearance and behavior from their ancestors. Their rapid rate of evolutionary change is exemplified by the cichlids of Lake Victoria. It dried out about 12,000 years ago but has since refilled and been totally recolonized, and many new forms have already evolved.

Cichlids are not all mouth brooders. Many are substrate brooders and nest in the open, the eggs being guarded by the male, female, or both parents, although male care predominates. Substrate brooding is considered the early evolutionary stage of reproduction in the cichlids; most species are monogamous, and the larvae are also guarded in the nest by both parents until they can swim well. When the eggs hatch, the remains of the yolk-sac nourish the larvae as they learn to swim. Nests can take several forms, from a simple depression made by removing sand from beneath an overhanging rock, or in a "cave" between rocks. These fish fan their eggs to ventilate them and increase the oxygen, and they suck them to remove wastes and dead eggs.

Most New World cichlids are substrate brooders; some may take their eggs in their mouths for short periods during the incubation cycle, but are not considered true mouth brooders. They lay many more eggs than the mouth brooders, as they are not restricted to the number they can hold in their mouths. They include many very popular aquarium fishes, including the red oscar (*Astronotus ocellatus*), which lays over 1,000 eggs on rocks that have been cleaned first of algae or debris. The hatchlings are then moved to pits where they are guarded by both parents, and they may sometimes cling to their parent's sides. The fire-mouth cichlid (*Thorichthys meeki*), another aquarium favorite, lays up to 500 eggs on rocks. In this species both parents also share the raising of the fry, which they protect in pits, moving them frequently to other pits. After the female orange cockatoo cichlid (*Apistogramma cacatuoides*) has laid her eggs in caves, the male enters to fertilize them but does not stay, and the female cares for the eggs and the hatchlings. The extreme development of parental care is shown by the discus (*Symphysodon discus*) of the Amazon River basin, the only known example of a fish actually feeding its young—with skin secretions from its own body. Other cichlids encourage their fry to eat by dropping food in front of them. This highly unusual caregiving behavior in cold-blooded animals also occurs in frogs that lay eggs for their tadpoles to eat, but is totally absent in the reptiles. The spectacular angel fish (*Pterophyllum scalare*), another cichlid of the Amazon Basin, which has a compressed body and long dorsal and anal fins, lays its eggs on a flat rock. Both parents guard them until they hatch, and clean them by sucking them into their mouths and then spitting them out.

Many African cichlids are also substrate brooders, either laying their eggs in the open or in the shelter of a cave. The genus *Julidochromis* contains several dwarf cichlids that are monogamous substrate brooders. Within the large genus *Neolamprologus,* one group—the shell-dwelling cichlids—live in old shells and spawn communally, with all the members of the group guarding the fry. Another group in the same genus, known as the fairy cichlids, vary in their parental care—some are polygamous and others monogamous—but they are all substrate spawners.

The kribensis or purple cichlid (*Pelvicachromis pulcher*), from southern Nigeria and Cameroon, deposits her eggs in an underwater cave and then cares for them by fanning them and sucking off the debris, while the male is outside guarding the territory. The butterfly cichlid (*Anomalachromis thomasi*), also of West Africa, lays its eggs on a cleaned stone and then moves the fry to a pit where both parents guard them. These substrate spawners lay lots of eggs, more than the mouth brooders but not the massive numbers of the oceanic egg-broadcasters.

The ultimate in egg-caring behavior occurs in the mouth brooders that hold their eggs and fry in their mouths. Many fish "incubate" their eggs in this way, and it is usually undertaken by the males except in the cichlids, where it is almost always the mothers' duty. The eggs are laid and fertilized by the male, and are then taken into the parent's mouth, where they complete their incubation. When birds and mammals incubate their eggs, they provide warmth for the development of the embryo; but cold-blooded fish cannot do this, and the temperature of the water in which they live is the incubation temperature, whether they hatch externally or are brooded in the mouth. It is a very restrictive form of reproduction and care, for the parent cannot eat while brooding and the number of eggs they can care for is limited by mouth size.

Mouth brooding is known to occur in nine fish families:

Cichlidae—the cichlids, with many mouth brooders, either paternal, maternal, or biparental, and others that guard their nests and eggs.

Osteoglossidae—the arowanas, large, bony-tongued, prehistoric-looking fish with thick scales. Some are paternal mouth brooders, others make nests for their eggs.

Ariidae—the sea catfishes of subtropical and tropical waters. They resemble freshwater catfish, and the male carries the eggs in his mouth.

Opistognathidae—the jaw fish; small, long fish with large heads and mouths, that live in burrows in the sand and hide when threatened. The males are mouth brooders.

Apogonidae—the cardinal fish, a group containing over 300 species living in the Atlantic, Indian, and Pacific oceans. The males are mouth brooders.

Belontiidae—the gouramis and other "labyrinth" fishes. Some species are paternal mouth brooders.

Bagridae—two members of this catfish family, *Phyllonemus filinemus* and *P. typus,* of Lake Tanganyika, are biparental mouth brooders.

Luciocephalidae—the pikeheads, perch-like fish that are all paternal mouth brooders.

Cyclopteridae—the lumpfish or lumpsuckers, some of which are paternal mouth brooders.

The African cichlids are the mouth brooders par excellence, and they practice the most widespread parental care in fish. There are perhaps 2,000 species in almost 200 genera, with most providing either female or biparental care. They live in the freshwater lakes and rivers of sub-Saharan Africa, but the most well-known aquarium species hail from Lakes Tanganyika, Nyasa (Malawi) and Victoria in East Africa, although the introduction of Nile perch into Lake Victoria has resulted in the extinction of many species there. Most are maternal mouth brooders, but they show a wide range of parental care of their offspring, with either one or both of the parents providing the care; and in the genus *Eretmodus,* it is shared. Most species are polygamous. Parental care by mouth brooding evolved among cichlids that lived in exposed places, where their eggs were more vulnerable to predation than the fish that occupied rocky habitats. Restricted by the number of offspring they can hold in their mouths, they lay fewer, but generally larger, eggs than the fish that lay their eggs on the substrate. When the eggs hatch, the larvae or fry remain in their parent's mouth for varying periods, and even after they have left they rush back to its safety when they are scared. Cichlids encourage their young to feed by picking up and dropping small pieces of vegetation in front of them, and they dig in the substrate to stir up buried organic particles.

Despite the safety that mouth brooding appears to offer, some cichlids have specialized in egg-eating and dash between the mating pair to steal their eggs as they are being laid and fertilized, and before a parent can scoop them up. Also, the Lake Tanganyika catfish (*Synodontis multipunctatus*) is a parasitic species that lays its eggs alongside a breeding pair of cichlids, usually selecting a pair of the "peacock group" *Aulonacara,* which then mouth brood the catfish eggs and eventually the fry.

Another group of mouth brooding cichlids, of the genus *Tilapia,* originated in Africa and the Middle East, but have been introduced widely and are a pest in several places, including Indonesia and Hawaii, and were purposely released in Alabama for sport fishing and in Californian waters to control weeds in irrigation channels. Tilapia, mainly the Nile tilapia (*Oreochromis niloticus*), are now farmed in many countries for food, and are an important commercial freshwater fish, a major protein resource in Malaysia, Indonesia, Thailand, Bangladesh, India, and several South American countries. Males excavate their nests in shallow water, and each then mates with several females, which lay their eggs in the nest, where they are immediately fertilized by the male. The females then take the eggs into their mouths, where they remain until they hatch. In the black-chinned tilapia (*Sarotherodon melanotherum*) the typical roles are reversed, and the males mouth brood the eggs and young. With their mouth partly open, they roll the eggs around to increase the oxygen flow. The fry also remain in their parent's mouth while they are absorbing their yolk sacs, and may even rush back to its safety after they are free swimming.

The mouth brooders are divided into two kinds. The ovophiles, or immediate mouth brooders, pick up their eggs as soon as they are laid and fertilized, and carry them in their mouths until they hatch. They then guard the fry also, letting them out to feed and allowing them back in when danger threatens, and they are apparently able to distinguish their own fry from others by their smell. The lemon yellow cichlid (*Labidochromis caeruleus*) of Lake Malawi and the Mozambique tilapia (*Oreochromis mossambicus*) of Lakes Tanganyika and Victoria are both ovophiles. Males of the latter species make spawning pits, in which a female lays up to 350 eggs. The male fertilizes them and leaves, and the female takes the eggs into her mouth, incubates them, and then holds the eventual fry, for a total of about four weeks. In some cichlids, fertilization of the eggs occurs after the females have picked them up. They either pick up sperm the males have deposited on a rock, or pick at "egg-spots" on the male's anal fin, that resemble eggs and thus prompt her "picking stimulus," and in doing so get sperm into their mouths. *Geophagus steindachneri* is one of these species, laying about 50 eggs on a flat rock, usually a few at a time, and then taking them into her mouth. The male then releases his milt or sperm onto the rock, and she takes that into her mouth also to fertilize the eggs. When the eggs hatch, after about two weeks, she lets them out to eat, but they rush back when alarmed, and do not leave the security of her mouth until they are about six weeks old.

Other mouth brooders, which initially care for their eggs on a rock or in a nest until they hatch and then take the fry into their mouths, are called larvophiles. They spawn in a protected place, such as a cave, and when the larvae hatch, they are sucked into the mouth. This form of brooding is also called delayed parental care. It is practiced by some members of the genera *Geophagus* and *Sarotherodon*. Another New World mouth brooder, *Satanoperca jurupari,* is a geophagous or earth-eating species, although it does not actually swallow the earth. It takes mouthfuls of substrate from the riverbed in northern South America, sifts out the organic particles, and spits out the rest. Laying up to 350 eggs on cleaned rocks, the female then takes them into her mouth and holds them for two weeks until they hatch.

There are also numerous marine mouth brooders. Some of the sea or salmon catfish (*Ariidae*)—a genus of about 120 species—brood the eggs in their mouths. They have a similar life history to salmon, their younger days being spent in fresh and brackish waters, after which they migrate to the sea to mature. Reaching a length of 24 inches (60 cm) when adult, the female lays about 60 marble-sized eggs that the male then broods in his mouth. Another marine mouth brooder is the yellow-headed or pearly jawfish (*Opistognathous aurifrons*), a popular home aquarium species. It spends most of its time peering out of its burrow in the sand, which it keeps open by periodically spitting out mouthfuls of sand. The male holds the fertilized eggs in his mouth and deposits them briefly when he needs to feed or to clean out his burrow. The fry swim off soon after they hatch. The Australian cardinal fish (*Apogon limensis*), a small, striped pink-and-black fish of the eastern coast of Australia that lives in rocky estuaries and reefs and is common in Sydney harbor, is also a mouth brooder. It carries its eggs and fry in its mouth and the fry dash back quickly to its safety when threatened. The eastern gobbleguts (*Vincentia novaehollandiae*), also of the coastal waters of eastern Australia, has a large oblique mouth that it crams with eggs.

Brooding eggs within their bodies also occurs in several fish, and the most unusual parental care arrangements are practiced by the members of the marine family *Syngnathidae*, which contains the seahorses and pipefish. Some pipefish are skin brooders, their eggs being either attached to the skin of the male's belly or actually embedded into the tissue. In a case of sex-role reversal, the male seahorse has an abdominal brood pouch with a small opening into which the female lays her eggs. He then fertilizes them and carries them until they hatch after an incubation period of about 18 days. When the babies are ready to leave, he forces them out through the same hole, and the female then lays another batch of eggs in the pouch. Seahorses generally form a pair bond for the breeding season.

In an even more bizarre arrangement, blind cave fish of the family *Amblyopsidae* incubate their eggs in their gill chambers. They are small cave-dwellers from the

Seahorse *A heavily "pregnant" seahorse. Bonding for life, an unusual situation in fish, seahorses also practice role-reversal, with the male caring for the eggs within his body. He has an abdominal brood pouch with a small opening into which the female lays her eggs. He then fertilizes them and carries them until they hatch and are released, whereupon the female lays another batch of eggs in his pouch.*
Photo: SF Photography, Shutterstock.com

southern and eastern United States, with vestigial eyes, no optic nerve, and no pigmentation, that can tolerate the low oxygen content of water in deep caves. The female southern cave fish (*Typhlichthys subterraneus*), a small pink, blind fish dwelling entirely in caves in the southeastern United States, incubates its eggs, and possibly broods its fry, in her gill chambers for a period of up to five months. Not surprisingly, this fish has a low reproductive capacity. But the most unusual care is provided by the humphead or nurseryfish (*Kurtus gulliveri*), from northern Australia and New Guinea, in which the male broods several hundred sticky eggs in a clump attached to a projection on his forehead. He is believed to collect them as the female lays them and then releases a cloud of sperm, through which he swims to fertilize them.

One other group of fish that must be included here, although they do not exhibit post-egg-laying care, are members of the cartilaginous class of fish (*Chondrichthyes*), which have improved the survival prospects of their offspring through pre-birth parental care. They are the egg-laying or oviparous sharks (one-quarter of all the sharks lay eggs) that include the leopard shark, cat sharks, bullheads, carpet sharks, swell sharks, and the dogfish. They lay only a few very large eggs, each enclosed in a horny, waterproof capsule that has long elastic tendrils from each corner. As she lays each egg, the female circles around seaweed to attach it, where it remains, unless broken free by storms, for its incubation, which may take almost one year. She has also provided each egg with a large yolk to nourish the embryo, so that it is quite well developed when it hatches.

Eggs are far more vulnerable than hatchlings, which can disperse and hide from predators; so the obvious answer to protecting the brood is to bypass the egg stage, which many fish have done, giving birth to live young. Fish were the first vertebrates to fertilize their eggs internally, and it is a far more common practice than in the amphibians that followed. In the sharks, the pelvic fins are modified for introducing sperm into the female, whereas in the top minnows (*Poecilia vittata*), the anal fin is modified for this purpose. With less risk to their young, the live-bearers produce few offspring at a time compared to the egg-layers, but a lot more energy is certainly expended in their development prior to birth. Live births result from fertilized eggs that develop internally, to the stage at which they are expelled as soon as they hatch, but parental care is unknown in any of these fish. Many familiar aquarium fish are live-bearers, including the guppies, mollies, and platies.

■ SOME OF THE SPECIES

Lemon Yellow Cichlid (*Labidochromis caeruleus*)

One of the mbuna or rock cichlids of Lake Malawi, this lovely species is a favorite aquarium fish, bright yellow with a long black-edged dorsal fin and black and yellow pelvic and anal fins. It reaches a length of 4 inches (10 cm), and it has a life span of 10 years. An omnivorous species, it feeds on aquatic invertebrates, especially tiny snails, plus vegetation, particularly algae. It lives along the rocky shores of the lake and also in the beds of water weed that grow in the shallows. The lemon yellow cichlid begins breeding at the age of six months, but only lays a few eggs then, increasing to about 20 eggs at each laying when she is older and larger. She lays her eggs on a rock, cleaned off for the purpose, and the male follows behind her and fertilizes them. She takes them into her mouth, then lays a few more eggs and the process is repeated. When she has finished laying the male leaves to find another gravid female. The brooding female does not eat for four weeks while the eggs are incubating, and then for a few days afterward until the fry have been released.

Blue Discus (*Symphysodon aequifasciatus*)

A cichlid, the blue discus has been an important fish in the aquarium trade for almost a century. It hails from the upper Amazon in the region of Leticia and the Rio Purus, where it lives in the clean, unpolluted waters of small tributaries and lakes. Typical of its genus, it has a disc-shaped or laterally compressed body, and grows to about 6 inches (15 cm) long. It has nine dark vertical bands that fade with age, and its body color ranges from yellowish brown to light turquoise, with deep blue fins, but three recognized subspecies add to the many variations in color. The blue discus is carnivorous, and its diet is mostly mosquito larvae and similar small aquatic invertebrates. It spawns on a slab of rock or a submerged log, which both parents clean off first. The female then lays up to 200 eggs, which she guards and oxygenates while the male guards the territory. The eggs hatch within two to three days, and the parents may nibble at the egg casings to help the larvae out. They then move them to a new site and attach them there with sticky filaments. After a few days the young follow their parents closely and feed upon secretions from the adult's skin, which is their sole food for the first week. Both male and female produce food for their young in this manner.

Oscar (*Astronotus ocellatus*)

One of the most popular aquarium fishes, the oscar is a cichlid from the upper Amazon and Rio Ucayali drainage basins in Peru and western Brazil, and is also established in the waterways of southern Florida. It is a large and stocky fish with an oval-shaped body, and it has extremely variable coloration, from olive-green to gray to chocolate mottled with red, and a very distinctive large black spot ringed with orange on either side of the tail peduncle—the main stem of the tail.

Oscars reach a length of 15 inches (38 cm) and weigh up to 40 ounces (1.1 kg). They are carnivorous and eat small fish, crayfish, and aquatic insects. At breeding time, the male and female clean off the surface of a flat rock and the female lays up to 2,000 eggs. After fertilizing them, the male stays nearby on guard and aggressively defends the eggs and the hatching fry. Oscars need a large aquarium and can only be kept with similar-sized fish; but they are notorious for disrupting their aquarium, uprooting and destroying the plants and rearranging the rocks.

Blue Tilapia (*Oreochromis aureus*)

This fish is often confused with the Nile perch (*Lates niloticus*) but is a much smaller species, growing to 22 inches (56 cm) long, and weighing 12 pounds (5.4 kg). The adult has a blue-gray color with a whitish belly, and its caudal and dorsal fins have reddish-pink borders. A native of northern Africa and the Middle East, it is believed to be the fish that the Bible records as feeding the multitudes. It has been introduced into Florida, where it is well established in freshwater and in brackish water along the western coast. It is a filter feeder that eats mainly algae, plus aquatic invertebrates. The name tilapia is actually generic for a group of cichlids that are important commercially, as they are farmed in several countries. This species, like all tilapia, is a nest-builder, and fertilization of the eggs occurs in the large nest cavity the male makes by scooping out sand with his mouth. He swims out to escort passing females to his nest, and fertilizes the eggs as soon as they are laid. The female then takes them into her mouth and swims off with them, mouth brooding them until they hatch, and for about three weeks afterward, when the fry return to her mouth if they feel threatened. The male continues to guard his nest and attract other females. This is the form of reproductive behavior known as polygyny—when a male mates with several females.

Black Bullhead (*Ameiurus melas*)

A medium-sized catfish from eastern North America, where it prefers slow-moving waters with muddy bottoms, the black bullhead has a high tolerance for turbid and low-oxygenated water. It reaches a maximum length of 18 inches (46 cm) and a weight of 35 ounces (1 kg). Small bullheads are preyed upon by larger fish, especially white bass, but they have sharp spines on the front of their pectoral and dorsal fins that can be locked to make them rigid, so larger individuals are hard to swallow. At spawning time, the female digs a nest in shallow water beneath vegetation or under an overhanging bank. When a male arrives, she releases some eggs at his prompting—he repeatedly nudges her head with his tail fin—and he fertilizes them as they are laid. She leaves when she has finished laying, and the male stays to guard the eggs and fan them to increase oxygen movement around them. After they hatch, he continues to watch over them, keeping them in a tight school and driving stragglers back to the pack.

North American Bluegill (*Lepomis macrochirus*)

The bluegill or sunfish resembles a perch and is a member of the sunfish family (*Centrarchidae*), which also contains several species of bass, and like them is a popular sport fish in eastern and central North America and the Great Lakes. It favors warmer waters with beds of weeds and submerged logs, and stays in the shallows unless the water becomes too hot in midsummer, when it seeks deeper and cooler levels but rarely goes deeper than 16 feet (5 m). When young, the bluegill eats algae and tiny aquatic creatures, graduating to small fish and insects as it matures, when it may be 15 inches (40 cm) long. It spawns in shallow water in early summer, and is a species that shows exclusive paternal care. Several males make a communal nest in the mud or sand and attract females who lay their eggs in the nest. The males then fertilize them, fan them to circulate oxygen around them, and stand guard over the eggs and then the hatchlings until they can swim properly.

Blue Gourami (*Trichogaster trichopterus*)

Also known as the three-spot gourami, this is a very popular aquarium fish, well-domesticated and now available in many color and pattern mutations. Its natural range is the fresh waters of Southeast Asia. It reaches a length of 5 inches (10 cm) and has a life span of about four years. The blue gourami is a bubble-nester and one of the labyrinth fish, possessing a special organ that allows them to breathe air at the surface, in addition to breathing through their gills. Spawning begins when the male makes a nest of saliva-coated bubbles, and then entices a female to it by displaying—flaring his fins and raising his tail. They then embrace beneath the nest, the male wrapping his body tightly around the female, and turning her on her side so that her eggs fall free. She may lay several thousand eggs, which he fertilizes as they are laid; and as they are lighter than water, they float up underneath the nest and lodge among the bubbles. She then leaves, and the male guards his nest until the eggs hatch, in about 36 hours, and then guards the fry until they are free-swimming.

Fringed Darter (*Etheostoma crossopterum*)

A small fish of the perch family, the fringed darter is about 3 inches (7.5 cm) long, and dark green in color. It lives in rocky-bottomed streams from Alabama to

Illinois. The male establishes his territory beneath a flat stone, and females arrive and attach their eggs to the underside of the stone; but they may also lay their eggs in more exposed positions on logs and other bottom debris. Several females lay their eggs in the same place until up to 2,000 eggs have been laid. The male fertilizes each batch as soon as it is laid and then guards them until they hatch after a seven-day incubation period, although this varies according to the water temperature. He fans them to circulate oxygen around them and sucks them to remove harmful bacteria and fungus. Darter eggs are popular food for many predators, and the male vigorously defends his eggs from minnows, salamanders, and crayfish, which are the major egg-stealers. It is thought that certain compounds in the mucus on the darter's skin inhibits the development of fungus and bacteria on the eggs.

Three-spined Stickleback (*Gasterostreus aculeatus*)

A small fish, about 4 inches (10 cm) long, the stickleback lives in fresh waters and brackish estuaries in the northern hemisphere. The species varies in color, but is generally olive-green above with bronze or silver below, and males develop a red breast during the breeding season. They have three dorsal spines, two pelvic spines, and an anal spine. Sticklebacks are carnivores and attempt to cannibalize the eggs of their conspecifics, so they have evolved parental care in which the male not only guards the eggs, but also makes this task easier by building a nest. This is not the typical depression scooped out of the sand that passes for a nest in so many species of fish, but an actual nest, made with filamentous algae and other pieces of vegetation glued together with excretions from the male's kidneys. It is attached to the substrate or to a plant, and is generally enclosed with an entrance at both ends. The female deposits her eggs in the nest and is then driven away, and the male then enters and fertilizes the eggs. The female may return again several times to lay more eggs. The male circulates water through the nest with his fins, and when the eggs are close to hatching seven days later, he breaks apart the cluster and scatters the eggs, and then chaperones the fry for several days.

Barbour's Sea Horse (*Hippocampus barbouri*)

This seahorse is a reef species of the western-central Pacific Ocean, a very popular fish in the aquarium trade. Seahorses have eyes on the sides of their heads for a wide field of view, a long snout, and a prehensile tail with which they can anchor themselves to coral or seaweed. They have tough skins and can change color like chameleons. They are poor swimmers, lacking a tail fin; propulsion is achieved with the large and transparent dorsal fin, and steering and stability with the small pectoral fins. Male seahorses are said to become pregnant, but they do not actually produce eggs in the normal manner. Their "pregnancy" is a very unusual form of parental care in which the female deposits her eggs into the male's abdominal brood pouch. They mate for life, and the female lays several hundred eggs directly into the male's pouch; he then fertilizes them and carries them while they incubate, which takes three to five weeks. He then releases the tiny seahorses, which are completely independent. The female lays more eggs in his pouch as soon as the previous brood has vacated it. In Southeast Asia, seahorses are dried and used medically to treat a wide range of disorders.

Lesser Spotted Dogfish (*Scyliorhinus canicula*)

A common small shark, with a cartilaginous skeleton, the lesser spotted dogfish lives in the northeastern Atlantic. It lays only about 10 eggs each month during the breeding season, which lasts from November to the following summer. For their protection, each egg is enclosed in a brown, almost transparent leathery capsule, known as a mermaid's purse, that is attached to rocks or seaweed by the long elastic tendrils extending from each corner. The female circles around the seaweed as she is laying to make sure the pouch is well secured. The eggs hatch after an incubation period of about nine months, the young fish being replicas of their parents, although only about 4 inches (10 cm) long. The dogfish is a bottom-feeder, reaching a length of about 30 inches (75 cm) and weighing 5 pounds (2.2 kg), and is sandy brown with small dark-brown spots. It has a very coarse skin that has been dried and used as sandpaper, and it is fished commercially, although it is sold by the more acceptable name of "rock salmon." Dogfish mate and fertilize their eggs internally, which is achieved through the use of specially adapted pelvic fins called claspers.

Shark Eggs *Embryos develop in shark eggs safe inside their horny, waterproof capsules in a marine laboratory. Many sharks lay eggs—just a few, quite large ones—and they have also further improved their chances of survival by incorporating long, elastic tendrils at the ends. As they lay the capsules, they circle around seaweed or rocks to ensure they are firmly attached.*
Photo: JoLin, Shutterstock.com

Asian Arowana (*Scleropages formosus*)

Also known as the dragon fish, this is a large and impressive species, a very popular aquarium fish in the Orient that is believed to symbolize luck, strength, and prosperity. It is a large and primitive fish, a member of the *Osteoglossidae* or bony-tongued fish, growing up to 39 inches (1 m) long, with large scales and bony plates covering its head. It has two sensory barbels protruding from its mouth. It is a long-lived fish that does not begin to breed until it is five years old, but it is now bred regularly and commercially, and exceptional specimens are worth several thousand dollars. The dragon fish is rare in the wild and is protected in most countries, where it occurs naturally in the rivers of Southeast Asia and Indonesia. It is a very carnivorous fish and eats practically anything of animal origin; it normally feeds on the surface but may leap out of the water to seize low-flying bats and birds. The dragon fish is a paternal mouth brooder, the male caring for up to 50 eggs in his mouth.

Ozark Cave Fish (*Amblyopsis rosae*)

This is a very rare fish with a small population, probably numbering no more than 2,000, that lives in pools in the limestone cave systems of Missouri, Arkansas, and Oklahoma. Blind and almost translucent, it reaches a length of only 2 inches (5 cm). Cave fish have lateral line receptor organs—sense organs that detect changes in the water's pressure and chemistry—with which they locate their food and detect potential predators, and they have special sensory papillae on their heads. They eat copepods (plankton), salamander larvae, tiny crayfish, and smaller cave fish, plus plant debris washed into the caves. They are usually found in pools beneath bat colonies and benefit, either directly or indirectly, from the bat's guano, which supports their prey. The risk to the cave fish's eggs from predation has resulted in a remarkable parental care strategy. After her eggs are fertilized, the female picks them up and then moves them into her gill chambers, where it is believed she holds them for several months while they incubate, giving the free-swimming larvae a better chance of survival.

2 Bubble-nesters and Back-packers

The amphibians began to evolve from fish with lobed fins about 390 million years ago, during the Devonian Period, and were the first tetrapods, or four-limbed animals, to walk on earth. They developed rudimentary lungs to breathe air, and their fins became more limb-like to support their weight as they walked on the lake bottom and then eventually came ashore, no doubt to escape predation and competition. This transformation took a very long time, however, as animals resembling modern amphibians did not appear until the Triassic Period, about 250 million years ago; the earliest known fossil with modern frog characteristics dates from the early Jurassic period, 190 million years ago. In the meantime, some of the early forms had diverged to become reptiles.

Amphibians are therefore the most primitive land-dwelling vertebrates, forming a living link between their gill-breathing ancestors and the lung-breathing reptiles and sharing several of the physiological characteristics of both. The class *Amphibia* contains three very different groups or orders of animals. There are about 5,300 species of tailless frogs and toads in the order *Anura;* approximately 550 species of tailed salamanders in the order *Caudata;* and 170 species of snakelike caecilians in the order *Gymnophiona.* Excluding the land-based caecilians, most amphibians are biphasic—having both a terrestrial and an aquatic phase at some time of their lives—but they can be considered terrestrial animals only in association with moisture.

Despite the passage of time since their evolution began, the amphibians have only been partially successful in colonizing the land and are still dependant upon water or moist places. They are either purely aquatic and spend all their lives in water, or they live in moist soil or in damp habitats usually close to water, even though some cannot swim. To acquire their life-supporting oxygen, many employ both gills and lungs, using gills in their larval stages, and in some species also throughout their lives, to extract oxygen from the water, and lungs for breathing terrestrially. Many salamanders still manage without lungs. In addition, the amphibians breathe through their skins, which have a rich supply of blood vessels and absorb oxygen from the water or the air. Like their ancestral fish and their descendants the reptiles, the amphibians are ectotherms, or cold-blooded animals, that cannot generate heat and control their body temperature like the endothermic mammals and birds, and must therefore select their environment to maintain their optimal temperature.

The first land animals did not totally give up their gills—which extract dissolved oxygen from the water—and during the course of their lives most of the modern amphibians pass through a gilled stage, but with some exceptions they lose them when they become adults. The exceptions are aquatic species that have gills throughout their lives, a condition known as neoteny or paedomorphism—the retention of larval characteristics into adulthood—that occurs in all the members of four salamander families, but not in any frogs or toads. But the amphibian's simple lungs did not solve the problem of acquiring sufficient oxygen, and they still rely on skin respiration, in which blood vessels near the skin's surface absorb oxygen for at least half their needs. Lacking gills and lungs, the lungless salamanders (*Plethodontidae*) are dependant on skin respiration, plus breathing through their mouth linings, and their thin skins have a rich supply of capillaries that almost reach the surface. The amphibians that need water, prefer it to be fresh. The tailless amphibians generally cannot tolerate salt water, as osmotic pressure would draw their body fluids through their semipermeable skins and dehydrate them. Some can live in brackish water, including the marine toad (*Bufo marinus*) and the clawed toad (*Xenopus laevis*). The crab-eating or salt-water frog (*Rana cancrivora*) probably has the highest tolerance for salinity, as it lives in coastal mangrove swamps. It secretes urea into its blood to bring the osmolarity close to that of sea water, so it does not lose water through its skin.

The amphibian skin contains glands that secrete either mucus or venom. Skin breathing requires a moist skin, and the mucus glands are important in the continual battle to prevent desiccation of the skin in the terrestrial species; whereas the granular or poison glands secrete substances that are highly toxic if swallowed and are quite corrosive on the skin. All the amphibians have poison glands, but their secretions vary greatly in toxicity. Toads, newts, and the poison arrow frogs produce toxins that can be fatal to other animals, including man, and the sheer size of the venom-producing glands (parotoid glands) of the marine toad—the lumps on their neck just behind the eye—make it a formidable foe. The glands are depressed when the toad is seized, and secrete venom in jets into the predator's mouth.

Most amphibians lay eggs that characteristically hatch into an aquatic larvae with gills and then metamorphose into young adults with air-breathing lungs.[1] Metamorphosis is the transformation of a larva into adult form, and in the amphibians it mostly means changing from an aquatic to a terrestrial existence. Legs are formed, and gills are replaced by lungs except in the species that remain totally aquatic. Their skins develop mucus glands to help keep them moist on land. Eyelids evolve and their eyes adapt for vision out of water. Frog and toad tadpoles lose their tails, but the salamanders keep theirs. From then on, more reliance must be placed on other senses as most species can no longer rely on the receptors along their sides, called lateral line organs, that sensed changes in water pressure or chemistry caused by predator or prey.

Unlike their fish ancestors, which must reproduce in water, many amphibians, in all three orders, have evolved to breed on land. But their eggs must be kept moist to prevent desiccation, as they still lay fish-like eggs with a gelatinous covering, lacking the shells and protective membranes of the amniotic eggs of the reptiles and birds. Amphibians are the only tetrapods that do not lay amniotic eggs. However, many species lay their eggs on land and have evolved unique methods of keeping them moist. Some terrestrial anurans bypass the tadpole stage, their eggs producing tiny replicas of the adults instead of larvae, and many salamanders and caecilians also give birth directly to living young. Most anurans practice external fertilization, which occurs during the body-clasping called amplexus, and ensures a high degree of paternal certainty. In contrast, all caecilians mate and fertilize their eggs internally and most salamanders have internal fertilization; the hellbender is an exception and fertilizes its eggs externally.

Care in the amphibians is associated with their reproductive mode. Members of all three orders employ the three typical forms found in the cold-blooded animals—

Coqui Frogs   *A native of Puerto Rico but now introduced into Hawaii and the Virgin Islands, the coqui frog is a terrestrial breeder and one of the few frogs to practice internal fertilization. During the eggs' three-week incubation period, the male "guards" them and may also hydrate them with urine. The female often also remains close by. The eggs have a large yolk and develop directly into tiny froglets, bypassing the tadpole stage.*
Photo: Courtesy Arnold H. Hara, University of Hawaii

oviparity (egg-laying), ovoviviparity (live-bearing without parental nourishment) and viviparity (live-bearing with nutrition from the parent), but the latter is very rare. Care is more closely associated with egg-laying rather than live-bearing, although not with the explosive breeding of temperate zone anurans in spring, forced on them by a short breeding season, when they lay many eggs to compensate for their lack of care and the expectedly high loss rate. The mating calls of thousands of frogs, ponds thick with frog spawn and then with shoals of black tadpoles, are the only times that amphibians are really obvious. The warm climate species are more secretive and less spectacular in their breeding efforts, but employ some of the most unusual, and even bizarre, behavior to safeguard their eggs and offspring. About 10 percent of the anurans care in some way for their eggs or offspring, or both.

The care the anurans provide includes creating foam nests to protect their eggs, physically guarding them against predation, moistening them with their damp bodies or with urine to prevent their desiccation, and transporting the eggs and tadpoles. In several species, the eggs become embedded in the parent's skin, where they develop and hatch as froglets. When the offspring (tadpoles or froglets) are also cared for, it usually follows the hatching of eggs that the parents have already been guarding. It also occurs when there is direct development within the egg, bypassing the tadpole stage, so that the young are born as miniature frogs. But only the anurans care for their offspring, as egg-attendance is the only form of parental care practiced by the salamanders and caecilians. About 20 percent of salamanders attend their eggs, especially the terrestrial Plethodontids. In the egg-laying caecilians (about 25 percent of the 170 species) the females of some species are known to guard their eggs.

Feeding their offspring is very unusual behavior for cold-blooded vertebrates, and with one exception—the discus fish—all the known examples are anurans. Despite the female's initial involvement, male care of the eggs and then the offspring, is more common. In the *Microhylidae,* a large family of frogs with about 350 species, paternal care predominates, and in a few species of poison arrow frogs (*Dendrobatidae*) both parents are involved. In the salamanders and caecilians, the females care for their eggs.

■ FROGS AND TOADS

Although scientifically the name "frog" applies only to members of the family *Ranidae,* and "toad" to species within the family *Bufonidae,* the members of all the other families are commonly called either frogs or toads, and collectively they are called anurans after their order *Anura.* The most obvious difference between them and the newts and salamanders is the fact that all anurans are tailless when they have metamorphosed from tadpoles. Frogs have smooth, moist skins and are more dependent on a wet or moist environment than the toads, and in addition to both terrestrial and aquatic forms, there are also arboreal species. They are more graceful in shape than toads, and have long hind legs for leaping.

Toads are stouter, often squat animals, with mostly dry, warty skins and short legs more suitable for hopping than leaping. Some have large parotoid glands, located behind the cranial crest, that produce very toxic secretions. They are more lethargic than frogs but are more adaptable, having evolved to cope with varying conditions involving considerable variations in humidity as well as temperature. There are purely aquatic species and terrestrial forms that can survive in very arid regions and require water only for breeding. Frogs and toads can be found on all the continents except Antarctica. Although Australia has no native true toads *Bufonidae,* its indigenous wildlife has suffered heavily from the depredations of the introduced marine toad (*Bufo marinus*).

Breeding

With just a few exceptions, the fertilization of frog and toad eggs occurs externally. Their mating involves the sexual embrace known as amplexus, in which the male hangs tightly to the female's back and fertilizes the eggs as they are laid. Amplexus may last for several days, or even weeks in some species, ensuring the eggs and sperm come into contact. Frogs with short legs, such as the African rain frogs (*Breviceps*), have a special secretion that "glues" the male temporarily to the female's back during amplexus. Internal fertilization is rare, and is practiced only by the coqui frog (*Eleutherodactylus coqui*) and seven species of toads in the genus *Nectophrynoides,* by apposition of the cloacas; and by two "tailed" frogs (*Ascaphus*), in which the tail is an extension of the cloaca and transmits sperm during mating.

There is a great range of reproductive systems in the anurans, despite the restrictions placed on them by external fertilization and amplexus plus the need to keep their eggs moist. In some species, both the eggs and the larvae are aquatic; others lay their eggs on land but then carry their tadpoles to water, whereas the eggs of other terrestrial species avoid dependency on water by hatching directly into froglets. Only a few species give birth to live young, in which the females retain the eggs in their bodies until they are ready to hatch. Amphibian eggs are protected only by a permeable gelatinous coat and rely for their water on external sources, unlike the amniotic eggs of the reptiles and birds that contain their own supply of fluids. Consequently, they must lay their eggs in water or in moist substrate to prevent desiccation. Amphibian eggs, like fish eggs, have only one inner membrane, called the embryonic membrane, whereas the amniotic eggs of the reptiles and birds have a shell and a series of inner membranes.

The three reproductive modes practiced by the frogs and toads are oviparity, ovoviviparity, and viviparity. Oviparity is the most common form of reproduction and the most familiar one, especially of the temperate climate explosive breeders that mate in water and have free-living larvae. Thousands of frogs congregate in spring—often even before the ice has completely melted—lay millions of eggs, and immediately abandon them. They provide the most obvious examples of amphibian reproduction, with jelly-like masses of eggs known as "frog spawn" covering the surface of ponds and marshes, and the less familiar strands or strings of eggs produced by toads, in both cases soon followed by shoals of shiny black tadpoles. But the eggs and larvae are left to their fate, their large numbers ensuring that a few survive the many dangers ahead, for there is heavy predation by other amphibians, fish, birds, snakes, and mammals. Surprisingly, with few exceptions, the anurans that provide parental care are egg-layers, and this form of reproduction is believed to be their ancestral or primitive condition.

The other two modes of reproduction are very rare in the anurans. Ovoviviparity, or placental viviparity as it is now called, is believed to occur in only three species. Two of these species are toads from central Africa, *Nectophrynoides viviparous* and *N. tornieri,* the other being the Puerto Rican or golden coqui (*Eleutherodactylus jasperi*)—the only member of the large New World family *Leptodactylidae* to give birth to live young. In these frogs, the eggs are retained within the female until the embryos are fully developed, bypassing the free-living tadpole stage, and she gives birth to froglets. However, the mother does not contribute any nutrients to the developing embryos, which are nourished entirely by their egg yolk. When frog's eggs develop "internally" after they are laid, such as in the stomach of the gastric brooding frog (*Rheobatrachus vitellinus*), or embedded in the back of the Surinam toad (*Pipa pipa*), they are considered to be cases of oviparity, not ovoviviparity, as their development occurs after they are laid.

True viviparity, when the larvae receive nourishment from the mother in addition to the egg yolk and are born as fully metamorphosed babies, occurs in only two anurans. They are the African live-bearing toads *Nectophrynoides liberiensis* and *N. occidentalis.* Their eggs are fertilized internally by apposition of the cloacas, and they have a long period of internal development akin to the gestation period of the placental mammals.

Despite the showy seasonal pond-breeders, many amphibians actually lay their eggs on land, where they must be kept moist at all times. The obvious development when water is unavailable is to bypass the tadpole stage; in about 20 percent of the frogs and toads the free-living tadpole stage is absent, as there is direct development within the egg to a metamorphosed juvenile—a tiny froglet. These species live mainly in the humid tropical regions, where they lay their eggs in the moist substrate of the forest floor, and most of these species "attend" their eggs, guarding them and keeping them moist. Examples of frogs lacking a larval stage include virtually all of the 600 species of the neotropical genus *Eleutherodactylus,* and the South Pacific island ground frogs of the genus *Platymantis.* Two of the three species of endemic New Zealand frogs *Leiopelma hamiltoni* and *L. archeyi,* and many species of *Microhylidae,* a family in which paternal care predominates, also lack a tadpole stage.

Care of the Eggs

Parental care in frogs is naturally impossible in the explosive breeders, for care requires a small number of eggs that can be protected. While this in turn requires more energy output, which is costly to the parents, it increases the chances of successful reproduction. Females must produce eggs with a much larger yolk, and therefore more nutrients, to feed the developing embryos. Consequently, these frogs lay fewer eggs than the explosive breeders. It is potentially costly for the caregiver also, as it is exposed to predators while watching over the eggs. Care of the eggs occurs mainly in equatorial regions, where moist and warm year-round breeding conditions prevail. It reduces predation in the vulnerable early stages of an amphibian's life, for many animals find frog's eggs very acceptable. Guarding the eggs is the most common form of care provided by the frogs and toads, and is practiced by about 10 percent of all species, mostly by the males. It protects them from some predators and from cannibalism by conspecifics. It is known to improve the hatching rate, so it is very important for species survival. When frogs were prevented from caring for their eggs in the cause of science, the mortality rate was much higher than normal.

The vulnerable nature of amphibian eggs to desiccation requires constant attention, and in addition to guarding them against predators, the parents crouch over them to keep them moist, and urinate on them to prevent their dehydration. Brooding behavior by the anurans is purely for their protection and has no relevance to birds incubating their eggs, for the amphibians cannot provide body warmth. If the care the anurans provide is vital, it is called obligate parental care; when the lack of care, such as the inability of the poison arrow frogs (*Dendrobatidae*) to carry their tadpoles to water, would result in their death. If it is not absolutely necessary all the time, it is called facultative parental care—such as in the gladiator frog (*Hyla rosenbergi*), in which the male protects the eggs if the risk is greater than normal due to the high density of other frogs in the area; otherwise he ignores them.

In addition to hydrating their eggs when necessary, if the offspring need water for their continued development, which is the case in practically all species, some frogs have solved that problem too by laying their eggs over water. When this behavior occurs in arboreal species, they lay their eggs on leaves over the water that collects in the axils—the angle formed between a leaf and stem. Madagascan frogs such as *Mantidactylus bicalcaratus* and *M. punctatus* lay their eggs on pandanus leaves just above their water-filled axils, into which the tadpoles drop when they hatch. Until the tadpoles hatch, the parents straddle the egg mass to protect it. In the Philippines, the toad *Pelophryne brevipes* also breeds in these tiny water "wells" that are called phytotelmata.

Making a nest for the eggs adds another measure of safety to overwater egg-laying. The African leaf-folding or banana frogs of the genus *Afrixalus* fold leaves over their fertilized eggs and then seal them together with a secretion from the oviduct, but they must stay nearby, ready to unseal them when the eggs begin to hatch. Most frogs, however, mainly tropical ones but also a few temperate species, have discovered that the ultimate way to protect their vulnerable eggs from desiccation and predation is to enclose them in a nest of foam. This is produced when a fluid secreted by the female during amplexus is beaten into a foam by her or the male. She then lays her eggs in these bubbles and the male fertilizes them, although in some species he ejects his sperm into the nest before the eggs are laid. The fluid whips up with the consistency of beaten egg whites and acts like an incubator for the eggs, keeping them moist and cool by reflecting the heat and guarding them against predation. It was recently discovered that the foam has a high protein content, which may contribute to the development of the embryos. The bubble-nest hardens on the outside, providing a moist microhabitat for the embryo's development. In some species the parents stay close to the nest, and may even cling to it, until the eggs hatch.

Anurans vary in their choice of site for their bubble-nests. Overhanging water is the favorite place, but others build on the bank of a pond or river, or in a burrow; and some even float their nests on the surface of the water, attached to vegetation to prevent them from drifting away. Bubble-nests are the preferred choice of the 280 species of tree frogs of the family *Rhacophoridae*. In Southeast Asia the gliding tree frogs of the genus *Rhacopherus* make a foamy nest attached to a branch over a pool or, more often, just over a puddle on the forest floor. The egg-jelly is beaten into a foam as the female lays her eggs and the male fertilizes them. The outer surface of the bubble nest hardens into a protective crust, providing a moist interior until the eggs hatch and the tadpoles drop out into the water, so it is important that the nest is strategically placed, or they would die.

In tropical America, many members of the huge family *Leptodactylidae,* containing about 1,270 species, also lay their eggs in foam nests that may be on the ground or in water. *Leptodactylus labyrinthicus,* of Brazil, nests on land near water, but only a few eggs are fertilized and the larvae then feed upon the other eggs in the nest. The Andean foam nesting tree frog (*Pleurodema cinerea*) is a very hardy species that makes its nest in water at elevations over 13,000 feet (4,000 m) in Bolivia. The toad-like tungara frog (*Physalaemus pustulosus*), of Central and South America, nests communally; several females lay their eggs and produce a jelly-like substance, which the males then beat into a foam with their hind legs. Another variation is practiced by the burrowing frog (*Leptodactylus fuscus*) of northwestern South America, which makes a hole about 2 inches (5 cm) deep in a mudbank at the waters' edge, and then makes its foam nest within the hole.

Foam nesters in Africa include the African foam-nest tree frog (*Chiromantis rufescens*) of the West African forests, which makes a foam nest over water to protect its 150 eggs. But the most spectacular species is the gray foam nesting frog (*Chiromantis xerampelina*), a cooperative nester in which up to a dozen frogs hang onto a branch and beat up a major foam nest for their eggs. Their larvae stay in the nest for two days after they hatch and wriggle to the periphery of the nest to breathe oxygen through the foam.

Foam nesters are rare in temperate climates, but in southeastern Australia and in Tasmania, two species make floating foam nests for their eggs. The eastern banjo frog (*Limnodynastes dumerilii*) or "pobbledonk," makes a large foam nest attached to a leaf in water, in which it lays up to 4,000 eggs; and the brown-striped marsh frog (*Limnodynastes peroni*) lays up to 1,000 eggs in a floating foam nest attached to water's-edge vegetation.

In the anurans, parental care of the eggs reaches its most extreme form in the species that carry them for the whole period of their development into tadpoles or tiny frogs. The most unusual is undoubtedly the practice of the gastric brooding frogs (*Rheobatrachus vitellinus*) and (*R. silus*) of Australia, both of which may already be extinct, as they have not been seen since the 1980s. The female swallows her 25 or so fertilized eggs and retains them, and the resultant tadpoles, in her stomach for the total six-week incubation and growth period, regurgitating them as tiny frogs. During this time she cannot eat, and suppresses the production of gastric acids and enzyme production to avoid digesting her brood.

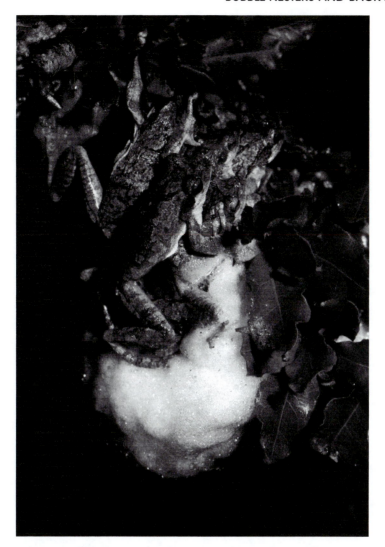

Foam-nesting frogs   *To protect their eggs from desiccation, many frogs make nests from a substance produced during egg-laying and beaten into a foam with their legs. These bubble nests are situated over water into which the tadpoles drop when they hatch. Some species, like the gray foam-nesting frog of eastern and southern Africa, are cooperative nesters in which several frogs hang onto a branch and beat up a major foam nest.*
Photo: Rod Preston-Mafham, Premaphotos Wildlife

A variation on this theme is practiced by the male Darwin's frog (*Rhinoderma darwini*), of central Chile and neighboring Argentina, who guards the eggs on the forest floor for two weeks. Just before they hatch, he takes them into his mouth and holds them in his vocal sac where they complete their metamorphosis, eventually appearing as tiny frogs. This is a case of oviparity (egg-laying), as the internal care occurs after the eggs are laid, whereas in ovoviviparous and viviparous species it occurs before. A related species, *Rhinoderma rufum,* also picks up his tadpoles when they hatch, but just carries them to water where he releases them.

Amplexus in the dorso-ventrally flattened Surinam toad (*Pipa pipa*) involves some very awkward-looking maneuvers, with the result that the fertilized eggs become attached to the female's back with a glue-like cloacal secretion. Within a few hours they sink into the highly vascularized skin on her back and become sealed in for the 75-day incubation period, finally emerging as tiny, flat toadlets. In the Andean marsupial frogs, of which there are 65 species in five genera including *Gastrotheca* and *Flectonotus,* the eggs and their developing embryos are protected in pouches on their mothers' backs. There they hatch as froglets in some species or as well-developed tadpoles in others, which then metamorphose in water. Getting the eggs into the mother's pouch involves some quick footwork by the male Riobamba marsupial frog (*Gastrotheca riobambae*) of Ecuador. As she lays her large yolk-rich eggs, the male in amplexus fertilizes them and catches each one with his feet. He then transfers them to a brood pouch on her back, where their development into froglets takes about three months. Another unusual case of carrying the eggs to protect them occurs in the midwife toad (*Alytes obstetricans*), in which the male wraps his mate's strings of eggs around his hindlimbs and carries them for several weeks, depositing them in a pool just before they hatch.

Care of the Offspring

Caring for the offspring is less common in the amphibians than caring for eggs, and it occurs only in the anurans, being unknown in salamanders and caecilians. Attending tadpoles or froglets that scatter upon hatching is obviously difficult, and parental care of the offspring generally involves controlling them by carrying them to water. This increases their offspring's survivorship and growth, improving the parent's breeding success. Generally, parental involvement is uncommon when development is completed in the egg—when the tadpole stage is bypassed and froglets are born. However, some species carry their offspring to water immediately after they are born; others have been observed to watch over their babies for a few days, and the male New Zealand frog (*Leiopelma hamiltoni*) carries his froglets, which still have their tails when they hatch, on his back for several days until their development is complete.

Carrying is mainly for the purpose of transporting the young from their place of hatching to another site more suitable for their continued development, and a number of frogs do this. Two species of Microhylid frogs, *Liophryne schlaginhaufeni* and *Sphenophryne cornuta,* both of New Guinea, lay about 25 eggs in moist debris on the forest floor. They then protect them and keep them moist during their month-long development. When the froglets hatch, they clamber onto their mother's back; she then makes her way after dark through the forest, on a journey that may last a week, dispersing the froglets, which hop off at various points along the way.

Many other anurans carry their offspring for much shorter journeys. The poison arrow frogs (*Dendrobatidae*) for example, are all terrestrial breeders that lay their eggs in moist soil, yet their young hatch as tadpoles, not as fully developed froglets. They must be moved as the forest floor is an unsuitable environment for them, so the adults carry them on their backs to water, usually the tiny pools that have collected in the axils of bromeliads and other plants high in the canopy. Both parents may carry the tadpoles, which climb onto their backs and hold on tight with their mouths or stick to a mucus-like secretion. In the green-and-black poison arrow frog (*Dendrobates auratus*), the male cares for the eggs and then the offspring. He protects and hydrates the eggs, and when they are due to hatch he crouches down, lowers his back, and makes a channel along which the tadpoles clamber and then attach themselves. He then carries them to water, usually up into the canopy to a bromeliad water well or phytotelmata. In the strawberry poison arrow frog (*Dendrobates pumilio*), the female usually cares for the offspring. She lays four or five eggs in a mass of jelly to keep them moist, and when they begin to hatch she steps into the jelly and frees the tadpoles. They climb onto her back, adhere to a mucus secretion, and are carried up to phytotelmata in the forest canopy, usually one per "pond" as they are cannibalistic. On the island of Borneo, the rough guardian frog (*Rana finchi*) and the smooth guardian frog (*Rana palavanensis*) live on the forest floor where the females attach their eggs to the underside of a leaf. The males guard them until they hatch and then carry the tadpoles on their backs to water.

The ultimate in amphibian parental care is feeding the young. It is the major difference in care, after the inability to provide warmth for their young, between the cold-blooded ectotherms and the warm-blooded birds and mammals. The anurans provide all the known examples except one (the discus fish) of parental feeding in the ectotherms. It occurs in numerous species in the families *Dendrobatidae, Hylidae, Rhacophoridae,* and *Mantellidae.* However, this is not a case of bringing worms or insects, bird-like, back to the young in the nest. The food that these frogs offer their young are eggs laid especially for this purpose. Several species of poison arrow frogs that are now regularly kept and bred by herpetologists, including *Dendrobates pumilio* and *D. speciosus,* feed trophic eggs to their young, which makes their captive breeding rather challenging. The Jamaican laughing frog (*Osteopilus brunneus*) lays a few eggs in a bromeliad phytotelmata, and after they have hatched, she returns to lay infertile eggs as food for the tadpoles. The spiny-headed tree frog (*Anotheca spinosa*) of Central America attaches her eggs to a leaf or stem above the water in a plant axil, and the larvae drop in as they hatch; the mother then returns to feed the tadpoles with food (trophic) eggs. Several *Chirixalus* species, a genus of frogs in the family *Rhacophoridae,* feed eggs to their young. In Eiffinger's tree frog (*Chirixalus eiffingeri*), the tadpoles have been observed jostling around their mother's cloaca in their eagerness to receive her eggs. The Madagascar poison frog (*Mantella laevigata*) lays its eggs above water in tree cavities and water-filled holes in bamboo. The larvae drop into the well when they hatch, and the mother returns to lay eggs as food for them.

■ SOME OF THE SPECIES

Marsupial Frog (*Gastrotheca riobambae*)

The marsupial frog, a nocturnal tree frog with mainly terrestrial habits, lives in Amazonian Peru and neighboring Brazil. Females are a bright pastel-green above and usually pale golden-brown below, with golden warts on their backs; and at 4 inches (10 cm) long they are double the size of the males, which are blotched dark-brown and pale golden-brown all over. The female marsupial frog has a pouch on her lower back in which she carries and incubates her eggs. As she lays her eggs while in amplexus, the male fertilizes them and pushes them into her pouch, aided by a secretion that he beats into a slippery foam with his legs. The tadpoles are eventually released from the pouch into water when they are about six weeks old, the mother holding the pouch open for them to get out. Adult marsupial frogs are not good swimmers and may drown in deep water.

Surinam Toad (*Pipa pipa*)

This toad is a most unusual, purely aquatic species from northern South America, in which the females reach a length of 6 inches (15 cm) and the males are a little smaller. It is grayish brown with pale underparts and has a rectangular shape, dorso-ventrally flattened body and limbs, and a pointed head. It has large flipper-like hind feet, but its forelimbs are short and unwebbed, and the fingers end in star-shaped sensory organs that replace the claws of typical frogs. Their toes are highly sensitive, and touch is the toads' major sense for locating prey. Surinam toads do not like turbulence, so they seek still waters where they live among the leaves on the muddy bottoms. There is no free-living larval stage in this species; for after mating and fertilizing the eggs, the male helps to attach them to the female's back, where they sink into her skin. She then carries them for their whole incubation period, usually about 20 weeks, and they hatch out as fully metamorphosed toadlets.

Mountain Chicken (*Leptodactylus fallax*)

Found only on the islands of Dominica and Montserrat in the West Indies, this large frog reaches a length of 8 inches (20 cm) and may weigh 2 pounds (900 g). It is a favored item of food and is known locally as the mountain chicken. Overhunting and loss of habitat exterminated it on other Caribbean islands, such as Martinique and Guadeloupe, and on Montserrat it is now very rare and has been further endangered by the recent volcanic activity there. In Dominica it has survived due to the protection of its mountain forests, where it lives in fast-flowing streams. The mountain chicken does not need water for its larval development, and the mating pair during amplexus make a foam nest in which the eggs are laid. Whereas most foam nesters place their nests over or near water, the mountain chicken creates its nest on land, usually in a burrow that may be 18 inches (46 cm) long, into which the male entices a female by trilling. The tadpoles are very large, up to 6 inches (15 cm) long, of which most is tail, and they are fed by the mother with unfertilized eggs. She stays close to the nest and aggressively defends it, while the male stands guard outside. The offspring leave the nest when they have metamorphosed into froglets.

Golden Poison Arrow Frog (*Phyllobates terribilis*)

This member of the *Dendrobatidae* lives in humid rain forest on the Pacific watershed of the Andes in Colombia. It is a uniform golden-yellow, which is aposematic or warning coloration to advise potential predators that it is an extremely toxic animal. Its is considered the most toxic of the poison arrow frogs, with enough batrachotoxin in one frog to kill many humans. Like the other members of its family, it is a small frog, reaching a length of only 2 inches (5 cm) when adult. The golden poison arrow frog breeds throughout the rainy season and lays its eggs in moist leaf litter on the forest floor. The male fertilizes them as they are laid, and then stays nearby and moistens them with urine when necessary to prevent their desiccation. When they are about to hatch, he kicks the egg mass with his hind feet to help free the tadpoles, which then crawl onto his back; there they adhere to a mucus secretion and are carried to the nearest water.

Coqui Frog (*Eleutherodactylus coqui*)

The coqui frog is a native of Puerto Rico, but it hid in house plants and was shipped with them, and is now established in the U.S. Virgin Islands and on the islands of Hawaii, Maui, and Oahu. The coqui is not considered a major nuisance yet in its new homes, unlike so many alien animals, but it does have a loud voice that many people find annoying. It is barely 1½ inches (4 cm) long, and its color varies considerably; its upper body may be gray, gray-green, or brownish-green, but the underparts are paler and are stippled with dark brown. It has pads on its toes like the tree frogs that allow it to cling to smooth surfaces. Like practically all the 700 members of the large genus *Eleutherodactylus,* the coqui frog breeds terrestrially, and within the egg the tadpole stage is bypassed and there is direct development into a froglet. This method of reproduction, called ovoviviparity, requires internal fertilization; and as the embryonic development takes longer, there must be sufficient energy within the egg to feed the larvae. The coqui's eggs are large, have a large yolk, and are laid on the ground. The male then guards them for three weeks until they hatch, and actually sits on them and hydrates them with urine when necessary. After they hatch, he may also watch over the froglets for a few days.

Darwin's Frog (*Rhinoderma darwinii*)

A native of the cool mountain streams in the temperate forests of southern Argentina and Chile, this tiny frog, just over 1 inch (2.5 cm) long, has an elongated, pointed snout and a triangular-shaped head. It varies in color from brown to green on its back, depending mainly on the background it is imitating, and its belly is darker with white and yellow spots. Its smooth skin has just a few warts. Darwin's frog has one of the most bizarre forms of parental care among the amphibians. After his mate lays up to 15 large eggs, with large yolks, the male guards them for their two-week incubation period. When he sees the tadpoles moving inside the eggs, he picks them up and holds them in his enlarged vocal sac, which extends down the sides of his body almost to his hind legs. They complete their growth in these "brood pouches," deriving their total energy from the remaining egg yolk. Their development to metamorphosis takes about six weeks, and froglets then hop out of his mouth.

This species is one of two in the genus *Rhinoderma*. The other—the Chile Darwin's frog (*R. rufum*)—picks up the tadpoles but carries them for only two weeks, he then releases them into water where they complete their development. This frog has not been seen for some time and may be extinct.

Southern Gastric Brooding Frog (*Rheobatrachus silus*)

A small species at just 2 inches (5 cm) long, from southeastern Queensland, this gastric brooding frog varies from greenish-brown to black on its back and has a creamy belly and yellow beneath its limbs. It was discovered only in 1973 and was considered quite plentiful, but then declined rapidly and has not been seen since 1981. Its disappearance coincided with that of many amphibians around the world, due in large part to pollution and climatic disturbances. Having avoided discovery until recently, there is hope that other hidden populations may still survive. Although it may be extinct, it must be included here as it is reported to have practiced an almost unbelievable form of parental care—certainly the most bizarre of the many strange systems the amphibians have evolved. An aquatic species, these frogs lived in rock pools in forested regions and apparently never moved more than a few feet from the water's edge. The female laid up to 25 eggs, and after they were fertilized, she swallowed them and held them in her stomach. When the eggs hatched, the tadpoles secreted hormones that inhibited the production of gastric acids that would have soon dissolved them, and the mother did not eat while carrying her young. The tadpoles stayed in her stomach for their whole development period, which lasted about six weeks, then metamorphosed and appeared in her mouth as froglets. Their diet during that period is unclear. Even among the incredibly diverse forms of parental care shown by the frogs, this seems an unlikely scenario.

A related species, the northern gastric brooding frog (*R. vitellinus*), was discovered in 1984 and has since also conveniently disappeared.

Javan Flying Frog (*Rhacopherus reinwardti*)

The Javan flying frog is an arboreal species from Southeast Asia and Indonesia that spends its whole adult life in trees, but its larval stage is aquatic. About 3.5 inches (9 cm) long, it is a very attractive frog, with a body color varying from light green to dark green, but its belly, toes, and the large webs between the toes are bright yellow, purple, orange, and black. These colors are very bright in the males, but are more subdued in the females, which also have smaller toe webs. Despite its name, it is a glider and not a flier. Opening its toes and spreading wide the webs, it can leap from the treetops and glide to the ground, from where it soon makes its way back to the canopy; except during the rainy season, when it makes a bubble nest over water, into which the tadpoles drop when they hatch. During amplexus, the female flying frog produces a fluid that is beaten into a foam, into which up to 800 eggs are laid and then fertilized by the male. Sometimes several pairs collaborate to make a very large nest in which all the females lay their eggs.

Midwife Toad (*Alytes obstetricans*)

The midwife toad, a plump and short-legged primitive toad in the family *Discoglossidae,* has a disc-like tongue and a warty gray skin. When adult it is about 2 inches (5 cm) long and, like the other members of its family, has vertical pupils. Although it lacks parotoid glands, the skin glands contain a very powerful toxin. A native of western Europe, it is terrestrial and lives in woods and gardens, but is rarely seen due to its nocturnal habits. It hides by day under logs or in crevices, or it may simply dig itself into moist sandy soil, where its singing while buried can be heard above ground. It is a very slow-moving toad, and unlike most anurans that breed in water, mating occurs on land, and then the male takes over the care of the eggs. He fertilizes the strings of eggs as they are laid, then wraps them around his hind legs and back and carries them for about three weeks before depositing them in water, where they hatch. A related species, the Mallorcan midwife toad (*A. muletensis*), was virtually exterminated before being saved by captive breeding programs.

New Zealand Native Frog (*Leiopelma hamiltoni*)

The most primitive of the world's frogs live in New Zealand, where they are not only endemic to the islands but are their only native frogs. Unfortunately, like so many animals there, they have been seriously affected by human occupation. There are four living species, three others having become extinct since the European settlement of the islands. Two of the survivors, *Leiopelma hamiltoni* and *L. archeyi,* lay their eggs in damp soil, and the male then crouches over them and guards them until they hatch into tiny froglets, although they still have tails. They climb onto

Midwife Toad *In one of the most unusual cases of parental care in the amphibians, the male midwife toad fertilizes the strings of eggs his mate has laid, and then wraps them around his hindlimbs. He carries them for several weeks, depositing them in a pool just before they hatch.*
Photo: Courtesy Jan Van Der Voort

their father's back and he carries them while they complete their development and have metamorphosed into tailless fogs. *Leiopelma hamiltoni* is the most plentiful of the four species, although by no means a common animal, occurring only in isolated pockets in the northern half of North Island. The others live respectively on Maud Island, Stephen Island, and in the Coromandel Range of North Island. They are terrestrial frogs that prefer moist forest habitat, where they hide under logs, and their terrestrial existence has resulted in their having only partially webbed feet. They reach a length of only 2 inches (5 cm) and are dark brown, and so are sometimes called brown frogs.

■ SALAMANDERS

The amphibian order *Urodela* contains the salamanders and the newts, which are known by those names but are also referred to collectively as salamanders. They are mainly northern, temperate-climate animals of both the Old and New Worlds, although they are especially well represented in North America, and a few species occur also in Central and South America. There are none in Australia.

Salamanders have long tails and cartilaginous pectoral girdles. Although they all have limbs, in some species these are tiny and useless, and in others they are present as forelimbs only. Their vertebrae range from 12 to 62. Salamanders have moist skins typical of the amphibians, and respiration occurs through their skins, and in some species via gills and lungs. The larvae of most species are aquatic and have external gills, and some retain these throughout their lives, but others have neither lungs nor gills and breathe only through their skins or mouths. The terrestrial plethodontids are lungless and breathe through their skin, whereas the aquatic adult amphiumas have no gills and breathe with their lungs.

Most salamanders have smooth and slimy skins due to the mucus secretions of their glands. Like the toads and some frogs, their skins also secrete toxins, some of them being extremely noxious and even fatal to humans if swallowed. Most species are bi-phasic, laying their eggs in water that hatch into aquatic larvae and later metamorphose into terrestrial or semiaquatic salamanders. Others, like the mudpuppy and hellbender, are permanently aquatic.

Unlike frog tadpoles that lose their tails when they metamorphose, salamander larvae grow legs but retain into adulthood their tails and, in some aquatic species, their gills also. Some even keep their gills throughout their lives and therefore breed in this "adult larval" condition, which is called neoteny or paedomorphosis. Life as a larvae can occur for a number of reasons. The most famous salamander to retain larval characteristics is the Mexican axolotl (*Ambystoma mexicanum*), in which the condition is genetically induced. Cold water prevents the release of the thyroid hormone that controls metamorphosis; and the olm, a blind eel-like salamander with a colorless body and tiny red gills, is prevented from metamorphosing by the cold water of the caverns in which it lives in Croatia. It also results from a deficiency of iodine in the environment, and there is a high rate of failure to mature in tiger salamanders (*Ambystoma tigrinum*) in some lakes in the western United States that lack this trace element. Neoteny is unique to the salamanders and occurs in nine of the ten families, and in the *Cryptobranchidae, Sirenidae, Amphiumidae,* and *Proteidae,* all the species are larvae for life.

Also, unlike the frogs and toads in which most species practice external fertilization, most salamanders fertilize their eggs internally, the exceptions being the Asiatic salamanders (*Hynobiidae*) and the giant salamanders and hellbenders (*Cryptobranchidae*). These account for only about 10 percent of the species, in which the eggs and sperm are deposited at the same time in typical frog and toad fashion, which is considered the primitive condition in the salamanders. Internal fertilization, however, does not result from coupling in the normal way, but as a result of males producing packets of sperm called spermatophores, that the females then pick up with the lips of their cloacas. A male may shed up to 100 spermatophores in a breeding season, and females can store sperm for up to two years. Parental care in the salamanders is confined to guarding the eggs by one or both parents, and none are known to provide care for their larvae.

The reproductive modes of the salamanders are exactly the same as those of the frogs and toads—oviparity, ovoviviparity, and viviparity, with oviparity being the most common. The oviparous species lay eggs that hatch outside the mother's body, the growing embryo being nourished by the egg yolk. The largest family—the lungless salamanders (*Plethodontidae*)—with about 250 mainly terrestrial species—are egg-layers, and they exhibit the most advanced care behavior of all salamanders. They find a cave or a burrow to deposit their eggs, in either a strip or a sticky clump; in many species, and possibly all of them, either the male or female attend the eggs. Their duties include the prevention of desiccation by resting their moist bodies on them, and possibly by urinating on them, like some frogs; and antifungal secretions from their skin are believed to inhibit the growth of fungi on the eggs. The Allegheny Mountain dusky salamander (*Desmognathus ochrophaeus*) may act as a foster parent for the eggs of other conspecifics, caring for them as well as her own.

Ovoviviparous salamanders fertilize their eggs internally, and these then develop in the mother's oviduct, being nourished entirely by the egg yolk, and the larvae break out of the membrane as the eggs are laid. As the larvae are aquatic, these species give birth directly into water, and there is no maternal care. Most of the many subspecies of the fire salamander (*Salamandra salamandra*) are ovoviviparous.

There are few viviparous salamanders, which must practice internal fertilization followed by a longer development period of the eggs and larvae within their mother. The larvae break out of the egg case and continue to grow inside her, feeding on the other eggs in the behavior known as embryophagy, and adding to this by scraping the mother's reproductive tract with their specialized teeth. The alpine salamander (*Salamandra atra*), is one of these species; only two babies are actually born, out of perhaps 25 eggs, after a gestation period of two years, and even longer at higher and therefore colder elevations. A race of the fire salamander, known as the yellow-striped fire salamander (*S. salamandra fastuosa*), is also viviparous. Ovoviviparity and viviparity are adaptations for life in harsh regions, especially high in mountains where the short summers are inadequate for normal larval development. Internal development in this manner, lasting two or even three years, protects the larvae during the winter, when of course their mothers would be dormant in hibernation anyway.

Salamanders' eggs are vulnerable to predation and must be guarded. The aquatic species, like the two-lined salamander (*Ensatina bislineata*), must keep fish and crayfish away from her 100 or so eggs that she attaches to a rock or to vegetation, and then guards them for their incubation period of 40 to 60 days. The eggs of terrestrial breeders are also vulnerable to frogs and slugs, as well as to their conspecifics, who consider them food. The eastern red-backed salamander (*Plethodon cinereus*) lays up to 16 eggs in cavities or in moist decaying wood, then watches over them for eight weeks, during which time she has to drive away other salamanders. In some species the young must be wary of their own parents, which may eat their offspring despite caring for the eggs for so long. But the very presence of some salamanders near their eggs is now known to be more than just physical protection against egg-eaters. The females, and occasionally the males, of the Monterey salamander (*Ensatina escholtzii*), guard their eggs for up to 12 weeks, moistening them with secretions that are believed to contain antibacterial and antifungal chemicals, a major hazard for eggs laid in damp conditions. Other salamanders known to aggressively defend their eggs are the spot-less paddle-tail newt (*Pachytriton labiatus*) and the spot-tailed warty newt (*Paramesotriton caudopunctatus*), both from China.

■ SOME OF THE SPECIES

Marbled Salamander (*Ambystoma opacum*)

The marbled salamander is one of the smaller species of mole salamanders, a terrestrial amphibian that is even less dependant than most on moist areas and can often be found living on quite dry hillsides. However, moisture is essential for reproduction, as the eggs are laid in a shallow depression on the forest floor. The mother attends them, protecting them from predators and preventing their dehydration, and her mucus may inhibit the growth of fungi. She cares for them until they hatch, which happens soon after the depression fills with water. The larvae are aquatic and breathe with their gills. They feed only at night, and when five months old they leave the water, lose their gills, and from then on live on land. The marbled salamander grows to about 4 inches (10 cm) long, and is basically black with extensive marbling on its back and sides, in the form of bands and blotches; these markings are usually gray in females and white in males. It lives in the eastern and southern United States from Massachusetts to Texas.

Alpine Salamander (*Salamandra atra*)

A native of the European Alps and the ranges of western Yugoslavia and Albania, the alpine salamander is 6½ inches (16 cm) long, including its tail. Most individuals are glossy black, but there are several subspecies, one of which—the golden alpine salamander (*S. a. aurorae*)—is golden-yellow on its back and tail and on the top of its head. The alpine salamander is nocturnal and lives in moist woodland, hiding beneath stones and logs during daylight. It hibernates in crevices and caves from October to March, often communally in a traditional hibernacula. On the forested mountain slopes it rarely has an opportunity to enter water, and this has resulted in the most unusual reproductive behavior of all the salamanders. The female produces a clutch of about 30 eggs, but instead of laying them she retains them in her body, where they develop. Only two complete their development, however, as they grow by eating the other eggs in the oviduct, and they eventually complete their metamorphosis within the mother and hatch as fully formed salamanders. This development takes two years to complete at lower elevations and three years at 5,000–6,000 feet (1,525–1,830 m).

Mudpuppy (*Necturus maculosus*)

The mudpuppy, so named because it and the related waterdogs are erroneously believed to bark, is a very long-bodied, eel-like salamander with tiny limbs. It lives in permanent rivers, lakes, and ponds in eastern North America—from southern Canada to the Tennessee River—where it reaches a length of about 12 inches (30 cm). It is grayish brown with scattered blackish spots, and its belly is gray with dark spots. Mudpuppy larvae are usually striped, with a dark line along the spine flanked on both sides by a yellow stripe. It is one of the salamanders that is a larva for life, like the more familiar axolotl; it retains its prominent external gills, which are usually colored maroon, and wavelike bird plumes in the water current. It is active at night when it searches for crayfish, insect larvae, snails, and fish in the dark waters; and in turn it is preyed upon by large fish and waterbirds. The mudpuppy lays up to 150 eggs in spring, usually attached to the underside of a submerged log or pond-bottom debris, and the mother guards them for their four-to-eight-week incubation period.

Hellbender (*Cryptobranchus allegheniensis*)

The rather grotesque hellbender is the sole New World representative of the family *Cryptobranchidae,* of which the others are the giant salamanders of China and Japan. It has a large flattened head and slimy, wrinkled folds of skin on the sides of its body that absorb oxygen from the water, and is either brown or black with indistinct spots. It is harmless despite its appearance. The hellbender's record length is 29 inches (74 cm), at which size it may weigh 5 pounds (2.2 kg). It is totally aquatic and nocturnal and lives in clear rivers and streams, usually in fast-flowing water, in the Appalachian and Ozark mountains of the eastern United States. It propels itself in the water with its short, tapering, and keeled tail. It hides under rocks and logs during the day, appearing at night to search for crayfish, fish, frogs, aquatic insect larvae, and snails. It has external gills only as a larva, and has small but muscular limbs. At breeding time—which usually commences in late summer—male hellbenders dig a large nest cavity beneath a log or rock, and gravid females are attracted to it, or may be driven to it by the males. Females lay two long strings of eggs in the nest, each containing about 200 eggs, that are coiled in a mass; and the male fertilizes them as they are laid. He then drives the female away, and stays in the nest to guard the eggs for their incubation period of approximately 70 days.

Japanese Giant Salamander (*Andreas japonicus*)

This large salamander reaches a length of 5 feet (1.5 m) and is the world's second-largest amphibian, slightly smaller than the Chinese giant salamander. It has a long body and long, broad tail, small limbs, a flat and wide head, and tiny eyes—with very poor vision. Its wrinkled skin is mottled black, gray, and cream. It lives in cold, fast-flowing mountain streams in northern Kyushu and western Honshu. Breeding commences in the fall when many salamanders congregate at traditional nest sites, where the males compete viciously for dominance and often die from their wounds. The successful males then make nests of mud and rocks under water, the entry to which is just large enough for access—a narrow horizontal tunnel about 36 inches (92 cm) long—leading to a nest cavity that usually has a sandy base. The male who built the nest occupies it and allows females to enter and lay their eggs, which he fertilizes, but surprisingly may also allow other males to enter and fertilize the eggs. Each female may lay up to 500 eggs that resemble threads of beads. The male then guards the eggs and attacks intruders. The incubation period is about four months, so the larvae begin to hatch just as winter is ending.

■ CAECILIANS

The caecilians are legless amphibians that have certain characteristics in common with the frogs and salamanders and are therefore believed to share a common ancestor. Externally, however, they are totally dissimilar, for they resemble worms or snakes rather than the other amphibians. They have long, limbless bodies, with up to 200 vertebrae, and thick, smooth skins, containing calcite scales, which reduce their flexibility. Their rounded and slightly pointed snout is an adaptation for burrowing, and tails are short or lacking in most species, hence the cloaca is near the end of the body. Their resemblance to large earthworms is enhanced by the annular grooves that form segments encircling their bodies, and some have toxic skin glands.

The caecilians are tropical animals, occurring in Central and South America, sub-Saharan Africa, and in southern Asia and Indonesia. They like moist soil and humid conditions. Most are gray, black, or purple; a few are yellow or pinkish, but there are some quite brightly colored ones like the banna caecilian (*Ichthyophis bannanicus*), one of the tailed species, which is purple above and yellow below. They lack ears and are deaf to airborne sounds, but are probably receptive to vibrations. Their name, derived from the Latin "caecus" meaning blind, is not totally descriptive of all species as many have small eyes, although in some they may be covered with skin and can only differentiate between light and dark. Their major sense organs are the sensitive tentacles between their eyes and nostrils, which have a chemosensory ability, and with which they locate their food. With few exceptions,[2] they are the only amphibians to have tentacles, which they can retract when they are not in use. They are all carnivorous, the smaller ones having specialized diets such as termite larvae, whereas the larger species eat worms, insects, fish, snakes, lizards, and frogs. They have many sharp teeth, which are for grasping, not chewing, as they swallow their food whole like snakes.

There are about 170 species of caecilians in six families. They are not a very well-known group of animals, as they have been poorly studied in part because of their life style; for although some species are totally aquatic, most caecilians are terrestrial and fossorial, spending practically their whole adult lives underground in a network of tunnels. The larvae of even the terrestrial species are aquatic before they metamorphose and go underground. One major family is the *Caeciliaidae,* with about 100 species of land and water forms, that are tailless and range in size from 4 inches (10 cm) to the large *Caecilia thompsoni,* at 48 inches (1.5 m) long and weighing 34 ounces (1 kg). The other large family is the *Typhlonectidae,* the members of which are aquatic and also tailless, and they give birth to larvae with external gills.

All the caecilians fertilize their eggs internally, the males having a penis-like organ for the transference of sperm to the females' cloaca. Seventy-five percent of the species give birth to live young, but it is unclear if they are ovoviviparous or viviparous. The embryos of some species are known to receive nourishment from their mother when they scrape cells off the oviduct wall with their tiny specialized teeth, and they possibly eat the other eggs inside the uterus. These species are therefore viviparous; but if any develop in the egg without such additional maternal care, and grow purely from the nutrients in the egg yolk, they must be considered ovoviviparous. However, they are born alive, and caecilians usually have up to 25 young. The remaining 25 percent of the caecilians lay eggs, and it is these species that provide parental care, for none of the live-bearers are known to do this.

The oviparous, or egg-laying species, find a water-filled hole in which they lay up to 50 eggs. These hatch into larvae, with gills and a tail, and feed on tiny plant and animal organisms. When they metamorphose, their gills are replaced with lungs and their skin thickens. Sensory tentacles grow and they come out of the water and go underground. With one exception, all caecilians have lungs when adult, and breathe also through the skin of their mouths. Like the snakes, their left lung is usually smaller than the right one, as a result of the elongation of their bodies.

Notes

1. Total metamorphosis within the egg occurs in some species, with the young emerging as fully-formed froglets, having bypassed the free-living tadpole stage.
2. One is the Surinam toad (*Pipa pipa*) that has short tentacles on its upper lip near the eyes.

3 Tender Loving Crocodiles

Reptiles were the first true land vertebrates, intermediate between the water-bound amphibians and the more recently evolved birds and mammals, and were the first vertebrates to become adapted to a completely terrestrial way of life. They began to evolve from the amphibians near the end of the Paleozoic Era about 350 million years ago, and became the dominant form of life during the following Mesozoic Era, the "Age of Reptiles." Those that survived into the Cenozoic Era (from 65 million years ago to the present day) gave rise to the almost 8,000 living species, just a small portion of this once huge class of vertebrates.

With the birds and mammals they are classified as amniotes (*Amniota*)—the higher vertebrates—which have a membrane called the amnion that envelopes the fetus during its embryonic development. They are adapted to a fully terrestrial existence or are secondarily aquatic; many still depend upon water for their food, protection, and breeding. They are the most advanced cold-blooded animals, their liberation from the water resulting from millions of years of evolution during which they developed lungs for breathing air throughout their life cycle, and eliminated the gill-breathing aquatic stage of most amphibians. They replaced semipermeable skins, which allowed respiration, with waterproof ones made of alpha keratin on the inner surface and the harder beta keratin on the outside, and so eliminated water loss through evaporation. They were the first vertebrates to lay eggs covered with a shell, in which the developing embryos are protected in fluid; consequently a watery medium is unnecessary for the development of their eggs, but humidity is still a requirement for egg incubation.

As they are still cold-blooded like their ancestors, however, reptiles cannot generate their own heat like the birds and the mammals, and are therefore at the mercy of their surroundings. Their muscle activity depends on chemical reactions that work faster in a warm body, so they are active when they are warm and sluggish when cold. This regulation of body temperature (thermoregulation) is achieved through a complex arrangement involving external heat sources, chemical reactions, hormone production, and behavior. With few exceptions, reptiles therefore cannot produce heat to incubate their eggs in the normal sense, and parental care involves providing a suitable site for their eggs and, in some species, guarding them and the offspring. They occur on all the continents except Antarctica and can be found in the hottest deserts and within the Arctic Circle. Most species live in the tropics, where the higher temperatures favor a cold-blooded lifestyle; but the temperate zones are home to a large number of lizards, snakes, and turtles, and the deserts there are especially rich in species.

The reptiles are all very familiar animals, recognizable on sight as snakes, lizards, tortoises, or crocodiles, the four major groups or orders within the class *Reptilia*. The lizards, snakes, and amphisbaenas, or worm-lizards, make up the largest order *Squamata*. The tuatara is the only living member of the order *Sphenodontia;* and the tortoises, terrapins, and turtles belong to the order *Testudines.* The order *Crocodilia* contains the crocodilians. The snakes and lizards are considered the most modern reptiles, because most of their speciation (the evolutionary process that gives rise to new species) has occurred in the past 60 million years, whereas tortoises have changed little in 100 million years. Like the amphibians, birds, the monotreme mammals (the platypus and echidnas), and most members of the order *Insectivora,* reptiles have a cloaca—the single chamber into which the digestive, urinary, and reproductive systems empty.

Reptiles were the first vertebrates to achieve the most successful adaptation for terrestrial life—the development of the amniotic egg—and were then followed in this great evolutionary advance by the dinosaurs, the birds, and then the mammals, where, with the exception of the monotremes, the amnion has become modified to form a placenta. Until then, water or moist places had been essential for the development of the vertebrate egg, and this new arrangement set the stage for the amniotes to take over the land. Many snakes and lizards have evolved to the next stage of reproduction, "incubating" their eggs within their bodies and giving birth to live young; a few have progressed even further and provide nourishment for their embryos within their bodies, in a similar manner to placental development in the higher mammals.

In the reptilian egg, unlike those of the fish and amphibians, cell division produces membranes, of which one, the amnion, forms a protective covering for the embryo that grows within the amniotic fluid. Membranes also enclose what eventually becomes the digestive tract and the cloaca for the expulsion of waste products. Amniotic eggs, bathed in their own fluid, do not have to be laid in wet situations, although they must not become dehydrated during their incubation.

Amniotic eggs must be fertilized before deposition of the eggshell, so this occurs internally in the reptiles as in the mammals and birds. Turtles and crocodiles only lay eggs, but snakes and lizards either produce eggs or give birth to living young. Snake eggs and those of most lizards have leathery, parchment-like "shells," whereas those of some turtles and the crocodilians are heavily calcified. When the eggs are laid, excluding the brooding pythons, they are dependent upon the warmth of the environment for their development. Inside the shell there is a protective membrane around the egg called the chorion, and next to this is the allantois, a sac connected to the embryo's abdomen that assists gas exchange, stores metabolic wastes resulting from the digestion of the egg yolk, and absorbs calcium from the shell for bone growth.

The development of the amniotic egg allowed the reptiles to lose their dependency upon water and colonize dry land. The eggshell protects the developing embryo from desiccation, and with less risk of predation, the reptiles could lay fewer eggs than their water-bound ancestors. Their young do not pass through a food-seeking larval stage followed by metamorphosis like most amphibians, and whether hatching from eggs or being born alive, the young are tiny replicas of the parents. However, the eggshell is porous and permits the movement of gases—oxygen in and carbon dioxide out—so it can equally lose water in dry conditions, which can kill the embryo. High humidity is therefore still necessary for the successful incubation of the amniotic egg.

American Alligator   *An American alligator guards its nest from raiders, especially raccoons. She lays up to 50 eggs, and then stays close to the nest for their whole 65-day incubation period. When the young alligators are ready to hatch, they alert her with high-pitched calls, and she opens the nest and carries the hatchlings to water in her mouth.*
Photo: Courtesy USGS

There are several theories about the evolution of the amniotic egg, but it is generally believed that it began developing in an early amphibian-like ancestor of the reptiles, about 320 million years ago, that still lived in the water but came ashore to lay its eggs. Evolution then favored eggs that were protected by a hard casing, and that allowed waste from the growing embryo to be stored in benign form within the egg. The development of the amniotic egg was dependent upon the evolutionary changes that permitted females to be fertilized internally; and it included the loss of an aquatic larval stage, allowing the developing amniotes to breed on land. With their horny nails, they could burrow to escape the heat of the day and bury their eggs, and eventually they completely lost the lateral line sensory system with which the fish and some amphibians detect movement in the water. Amniotic eggs are laid by the reptiles, birds, and the monotreme mammals. In the placental mammals, the development of an amniotic membrane allows nutrients and wastes to pass between the embryo and the mother. The laying of shelled eggs, however, did not eliminate parental care in the reptiles, although the members of each order generally restrict their involvement to just one aspect of care—either finding or preparing a site for their eggs, or guarding them, and in just a few species caring for their offspring.

Reptiles employ all three reproductive modes found in the cold-blooded vertebrates: oviparity, or egg-laying; ovoviviparity, or unassisted embryonic development within the mother; and viviparity, when the mother provides nourishment for the embryos. As in the amphibians, oviparity is by far the most common form, but in the reptiles there is greater development towards live-bearing, with 20 percent of species being ovoviviparous, compared to the few amphibians that give birth in this manner. True viviparity is just as rare in reptiles as in the amphibians.

In most oviparous species, the female lays fertilized eggs with undeveloped embryos, and development takes place after laying. This type of reproduction occurs in all the crocodilians, turtles, and in most lizards and snakes. In some reptiles (*Anolis* and *Elaphe* species), the development of the embryo may begin before the egg is actually laid, as they retain their eggs for part, and sometimes even half, of their embryonic development period before laying them, which is considered to be evolving ovoviviparity. An unusual aspect of the reproduction of many egg-laying species is the determination of sex by the egg-incubation temperature, as they do not have sex chromosomes so it cannot be determined genetically; instead their gonads are thermosensitive. Although their lifestyle may be totally aquatic, and they may even mate in the water, all egg-laying reptiles deposit their eggs on land, as immersion in water for more than a few hours is fatal to the embryo. Parental care is practiced by many of these species, most prior to or during the process of laying their eggs, some during their incubation, and a few after their young have hatched. Crocodilians may attend their hatchlings for several months, whereas parental care of the young by the skinks lasts only for a few days and seems mainly concerned with protecting them from infanticide by other adult skinks.

Ovoviviparity is the mode of reproduction in which the embryos develop within eggs that remain in the mother until they hatch or are about to hatch. They are born fully developed although usually still enclosed in the amniotic membrane, out of which they immediately escape; so this is considered live-bearing. The embryos are nourished entirely by the egg yolk, and do not receive any nutrients from their mother, but there is some gas exchange between mother and embryo through the amniotic membrane. Twenty percent of all snakes, lizards, and amphisbaenas are ovoviviparous, including all the boas and vipers, and some skinks. In the live-bearing reptiles, the hatchlings are independent at birth, and the Solomon Island skink (*Corucia zebrata*) is believed to be the only live-bearing species to care for its young.

Practically all live births in the reptiles result from ovoviviparity, and examples of true viviparity—in which the mother practices matrotrophy and nourishes her young—are quite rare. It is the most advanced form of parental care in cold-blooded vertebrates, but it has apparently evolved and regressed many times during the evolution of the snakes and lizards. It is believed to have evolved in response to cold environments and short summers, when retention within their mother allowed the young to develop over two years. Garter snakes and some baby skinks rely almost entirely upon nourishment from their mothers, but further studies are likely to reveal that this mode is more common. However, it has recently become fashionable to consider all live births in reptiles as cases of viviparity—whether the embryos are nourished by the mother or by the egg yolk—thus making the term ovoviviparity redundant.

Live-bearing has both disadvantages (for the mother) and advantages (for the offspring). With her oviduct full of developing embryos, the female is slowed down—both for hunting and when being hunted. She loses more fat producing babies than by simply laying eggs, which can then affect her future breeding prospects. Her babies, however, are safer inside her than eggs are outside and at the mercy of predators. Also, they are more advanced at birth and better able to take care of themselves.

Although certain reptilian behavior is clearly parental care, such as pythons incubating their eggs and crocodiles guarding their nest and then caring for their young, other aspects of their breeding behavior should also be considered parental care. These include the actions taken by many turtles, lizards, and snakes to bury or hide their eggs, protecting them from predation and ensuring the temperature and humidity of the environment is appropriate for their incubation. Although some captive reptiles have scattered their eggs around their cages, this probably results from the unnatural conditions, for no wild reptile is known to lay its eggs in such a haphazard manner. They choose their nest sites carefully, and in many species actually prepare the site to receive the eggs. Sea turtles return to traditional beaches to lay their eggs; tortoises urinate on their eggs to increase the humidity in the nest hole, Galapagos land iguanas crawl several miles to traditional, volcanically heated nesting sites, and snakes find a pile of decomposing leaf litter or a farmyard manure pile to provide the right conditions for their eggs.

After laying their eggs, few reptiles show any interest in providing further care. Among the snakes, the pythons and a few other species coil around their clutch to guard it and, in the pythons at least, raise their body temperature by muscle contraction to assist incubation. Just one species, the king cobra, actually makes a nest and then aggressively defends it until the eggs hatch. The crocodiles, however, despite their size and aggressive nature, are dedicated parents and show the most highly developed parental care of all reptiles. An armored monster weighing almost one ton (and in many species its mate also) guards the nest against predators and then gently carries the tiny hatchlings in its huge jaws to the water, assisting those that have difficulty escaping from the eggshells. Reptiles do not bring food to their young in the manner of the mammals, birds, and some amphibians.

Most reptiles are solitary creatures, and after mating they separate and continue their independent lives. Consequently, the male is likely to be some distance away when the female gives birth or lays her eggs, weeks or even months later. Parental care is therefore the mother's responsibility in most reptiles. Exceptions include the crocodiles, where the bond between the pair is more lasting in at least eight species, and both parents may guard the eggs and the young. The king cobra is the only snake in which parental duties are shared, and there are a few skinks and perhaps some geckos in which the male is believed to assist in guarding the eggs.

■ TURTLES

The aquatic turtles and terrapins and the terrestrial tortoises and box turtles are referred to collectively as turtles in North America, whereas in Europe the name turtle is normally used only for aquatic species. They are all members of the order *Testudinata,* which for many years was known as *Chelonia,* so they are often still called chelonians. Turtles are very primitive reptiles, having changed little in the last 100 million years. They generally require a warmer climate than the other reptiles, excluding the crocodilians, and do not range as far into the higher latitudes of the northern hemisphere as the snakes and lizards.

The turtles' most obvious characteristic is their "external skeleton," comprising a bony upper shell called a carapace and a lower shell or plastron. The carapace is a single shell, with its bony plates fused to the ribs and the trunk vertebrae; but in some species the plastron is segmented and hinged, allowing the turtle to withdraw its head, legs, and tail and close up tightly. Their shells vary in shape, some turtles having a flattened, pancake-like carapace, while others are quite high-domed or heavily ridged. The individual plates on the carapace are called scutes. Unlike the snakes that shed their whole skin at one time, shedding in the chelonians is a continual process with pieces of the scutes flaking off. Turtles vary in size from freshwater species just 3 inches (8 cm) long, to the giant land tortoises of the Galapagos and Aldabra islands that can weigh 650 pounds (295 kg), and the huge marine leatherback turtles, in which the record weight is 2,019 pounds (916 kg). None have teeth, but instead have powerful cutting and crushing jaws, covered with a horny sheath similar to a bird's beak, with a sharp edge in some species and a serrated one in others.

All the turtles lay eggs, which vary in their covering, having either a soft and flexible "skin" or a rigid, calcified shell. For example the box turtles' eggshells are thin-walled and flexible, whereas those of the Mediterranean tortoises have thick shells of hard calcium carbonate; but all are considered shelled eggs. The laying of shelled eggs did not eliminate the need for parental care in the turtles, however, and they all ensure their eggs are placed in the most suitable environment for their development. To lay their eggs, the marine turtles swim thousands of miles back to the beaches where they themselves hatched. The Galapagos tortoises urinate on the soil to soften it while digging their nest holes, and the desert tortoise actually urinates in the base of her nest hole and then on the eggs before filling it, to increase the nest's humidity. In the northern aquatic species like the painted turtle, selecting a suitable nest site may mean traveling some distance from her pond. With the exception of the desert tortoise, once seen defending her nest site against a Gila monster, when they have laid their eggs the turtles take no further part in the reproductive process. Parental care of the young is unknown in the turtles, and their hatchlings must make their own way in the world.

There are about 300 species of turtles, and all of them—the truly aquatic ones, the land tortoises, and the semiaquatic terrapins—lay their eggs on land and without exception they bury them, a typical reptilian trait, both for their protection from predators and from the sun. They select a location where the substrate is moist, for their shells are permeable and must absorb moisture during their incubation; but lengthy submersion in water is fatal to the embryo, so they are safe unless abnormally high water levels flood their nests. Similarly, they must dig their nest deep enough to prevent it drying out, for that would also kill the embryos. As the breeders of captive turtles know, to incubate their eggs successfully requires conditions of high humidity, generally near 95 percent; and when artificially incubating turtle eggs, they must be nestled in damp vermiculite or sphagnum moss in the incubator. Providing the correct temperature is also important, for the eggs of these cold-blooded animals, unlike those of the northern fish and amphibians, need high temperatures to develop, usually between 80°F and 85°F (26.7°C and 29.5C°).

An interesting aspect of turtle development is that the sex of the hatchlings is influenced by the incubation temperature because the embryos do not have sex chromosomes, so sex cannot be determined genetically. This is known as "temperature dependant sex determination." It occurs in many reptiles and is common in turtles, in both freshwater and marine species, because the embryo's gonads are sensitive to temperature during their development and are said to be thermosensitive. In the pond turtles and box turtles (*Emydidae*), for example, incubation temperatures above 78.8°F (26°C) produce mostly female hatchlings; whereas below it most babies are males. The ability to control the sex of baby turtles in this way has great significance in the captive breeding of these animals and therefore for conservation.

■ SOME OF THE SPECIES

Desert Tortoise (*Gopherus agassizi*)

This tortoise lives in the arid lands of the southeastern United States and neighboring Mexico, where ground temperatures may reach 140°F (60°C) in midsummer, and then must hibernate for the winter, so it is superbly adapted to combat harsh conditions. It may spend 90 percent of the year underground in its burrows, which average 36 inches (92 cm) deep and may be 10 feet (3 m) long, and can survive without drinking for up to one year. It is most active in spring and early summer when the vegetation grows after the spring rains, as it eats grasses, forbs, and the leaves and flowers of low shrubs. By early July it becomes dormant for the rest of the summer and may then continue straight into hibernation for the winter. The desert tortoise reaches a length of 15 inches (38 cm), and has a high domed shell or carapace that is usually horn-colored or brown, with prominent growth lines and yellowish centers to the scutes. Its head is small, its hindlimbs are cylindrical, and it has large scales on its forelimbs. The sexes are similar, except for the male's longer tail, and slightly concave plastron or lower shell. This tortoise nests in May and June, and digs two or three nest holes about 8 inches (20 cm) deep with her hind feet. She lays up to eight eggs in each nest, urinating in the nest first, then on the eggs, and then again on the surface afterward, in what is believed to be an effort to raise the humidity of the sand and possibly to deter predators. In what is the only record of a turtle providing parental care after laying her eggs, a female desert tortoise was seen on two occasions to vigorously defend her nest against a Gila monster.

Painted Turtle (*Chrysemys picta*)

The common North American freshwater species, the painted turtle ranges across the continent from coast to coast, and from the Mississippi delta north into southern Canada. About 8 inches (20 cm) in length, it has a low, smooth, and unkeeled carapace, usually dark green with red markings on the marginal scutes. Its plastron is yellowish orange with a long gray blotch down the middle, and the head and neck are distinctively striped with red, yellow, or orange lines. The painted turtle prefers the still or sluggish water of ponds, lakes, marshes, ditches, and prairie sloughs with muddy bottoms and profuse plant growth, and it basks on logs and mudbanks, often communally. It is omnivorous and eats aquatic plants and insects, crayfish, frogs, tadpoles, and snails, and also scavenges on carrion.

The painted turtle's clutch of eight eggs is laid in midsummer, usually in a hole scooped out of the south-facing sandy bank of a pond or lake, and is left to incubate by radiation. Many females travel some distance from their pond in search of a nest site, often crossing roads to find a suitable place with loose soil or sand and direct sunlight. The nest is flask-shaped, and after laying their eggs they cover them and return to their pond to hibernate. The eggs take about 10 weeks to hatch, and by then it is too cold for the hatchlings to be active, so they hibernate where they hatch, surviving until spring because they can produce cyroprotectants or "antifreeze" compounds that protect the cell contents from freezing.

Leatherback Turtle (*Dermochelys coriacea*)

The largest living turtle, the very distinctive leatherback is the only member of the family *Dermochelys*. Unlike the hard shells of the other marine turtles, its carapace is smooth, rubbery and slightly flexible. The leatherback is mainly an animal of the tropical seas, including the Atlantic, Pacific and Indian Oceans, but it has ranged north almost to the Arctic Circle in the Atlantic. The largest specimen on record weighed 2,019 pounds (916 kg) and measured 8 feet (2.4 m) long, and was stranded on a beach in Wales in 1988. It is now endangered, down to 25,000 adult females from 115,000 in 1988 and still declining rapidly although protected, due to accidental capture in fishing nets, egg collecting, and ingesting plastic debris

Leatherback Turtle *A leatherback turtle lays her eggs in a hole dug deep into the sand of a tropical beach. The largest of the turtles, weighing up to 2,019 pounds (916 kg), the female swims thousands of miles to lay her eggs, usually on the beach where she hatched many years before. A clear case of parental care before egg-laying.*
Photo: Lynsey Allen, Shutterstock.com

that it mistakes for the jellyfish that form its diet. After outliving the dinosaurs by millions of years, this ancient creature has been brought to the verge of extinction in just two centuries.

The leatherback nests on average eight times each season, laying about 80 fertilized eggs the size of billiard balls on a traditional nesting beach in the tropics or subtropics, after probably swimming several thousand miles to get there. She may change her nesting beach, from year to year, unlike the other sea turtles. Cleaning away any debris, she digs a pit for her body to rest in and then digs the deeper pit that serves as the nest hole. After laying her clutch, she refills the hole with sand, packing it down as she does, and then refills the body pit. To fool potential nest robbers, she spreads and levels the sand to camouflage the site, and then makes several false nests nearby. If left unmolested, the eggs hatch in 65 days and the tiny turtles must then run the gauntlet to the sea.

■ LIZARDS

Lizards are considered the most successful of the modern reptiles. They have colonized all the continents except Antarctica and live in a wide variety of habitats from the hottest deserts to tundra within the Arctic Circle. The are certainly the most plentiful reptiles, with about 4,450 living species, and are close relatives of the snakes and members of the same order, *Squamata*. The presence of limbs generally distinguishes them from the snakes, although there are actually many legless species that resemble snakes.

Lizards have hinged and flexible scaly skins, similar to those of the snakes. Their scales, composed of thickened epidermis, may be very smooth and arranged in rows as in the snakes, or may be ridged, spiny or keeled, and irregular. Even the legless species can be easily identified as lizards because they have eyelids, which snakes lack. In most species the lids are closable, but some lizards have transparent fused eyelids like the snakes, and a few have a transparent bottom and can still see when their eyes are closed. Some of the burrowing species are blind. Lizards lack both outer ears and external ears; and their sense of hearing, at least of airborne vibrations, begins at the tympanic membrane or eardrum, visible on the surface of the head, that covers the middle ear cavity. Ground vibrations are picked up by bones in the jaws that vibrate other bones in the middle ear, which pass the information to the inner ear, from where it is transmitted to the brain along the auditory nerves.

Most lizards are oviparous and lay eggs with tough, leathery, or parchment-like shells with the exception of the geckos, whose eggs contain calcium carbonate and are said to be calcareous. Although geckos' eggs are soft and pliable when laid, and stick to the surface on which they are laid, they soon harden and cannot be moved without cracking the shell. Practically all the live-bearing species are considered ovoviviparous, as the eggs are retained by the mother until they are about to hatch, and during their development they are nourished entirely by the egg yolk. Few lizards are viviparous, in which the embryos gain nutrients from their mother.

The egg-laying lizards include the monitors, anoles, most of the geckos, skinks, and chameleons, some iguanas, lacertids such as the wall lizard and sand lizard, the Gila monster, and the tuatara. Few species brood their eggs, so parental care in the egg-layers is restricted mainly to providing a site suitable for their development. Most bury their eggs, or at least hide them under logs or rocks, in crevices or old rodent holes, and some monitor lizards lay theirs in termite mounds.

The most extreme case of pre-laying parental care in lizards is shown by the land iguanas of Isla Fernandina in the Galapagos Islands. They crawl from their feeding grounds on the coast up to 6 miles (9.6 km) to their egg-laying sites on the rim of La Cumbre volcano, at 4,900 feet (1,495 m) above sea level, and may even go down the caldera's steep walls to the base, 3,000 feet (900 m) below, to lay their eggs in the heated volcanic sand. Unfortunately, their hatchlings must make the same return journey, running the gauntlet with snakes and hawks waiting for them. Lizards that do not bury their eggs include the geckos; some lay their eggs in the axil of a leaf or inside a cracked bamboo; others glue them to the underside of a leaf, and the tokay gecko (*Gecko gecko*) fixes her eggs to a wall out of the direct sunlight. In several lizards, including the tuatara, the incubation temperature is known to determine the sex of the offspring.

Although most reptiles reproduce sexually, many lizards give birth asexually by the process of parthenogenesis or "virgin birth." In this unusual form of reproduction, females produce eggs that develop without being fertilized, the ova having the same chromosome number as the mother. Members of six lizard families are known to do this, including several geckos, whiptails, and racerunners.

Few lizards care for their eggs; most species abandon them as soon as they have covered the nest hole. Skinks are the most protective species, and several brood their eggs, actually curling around or over them, and some geckos guard their eggs, staying close to deter predators. But in practically all cases it is the females' responsibility. There are few accounts of male lizards guarding, or appearing to guard, the eggs. Some male skinks have remained near the eggs, and a male tokay gecko stayed close to eggs attached to a wall during their incubation and may have been guarding them. Examples of skinks that brood their eggs are the sand skink (*Neoseps reynoldi*), the western skink (*Eumeces skiltonius*), and the broad-headed skink (*Eumeces laticeps*). The guarding female lunges at aggressors with an open mouth, and in the species that cover their eggs the mothers have been observed re-covering any that became exposed. Even in those that guard their eggs, care of the hatchlings is not highly developed. The young of some species stay near their mother for the first few days of their life, and there are a few accounts of skinks protecting their young from attacks by other skinks.

From simply laying fertilized eggs, the next evolutionary advance was to retain them in the oviduct and give birth to live young, and many species are obviously in the process of doing this. In one Australian lizard, both egg-laying and live-births actually occur, albeit in a different environment. On mainland Australia, Bougainville's skink (*Lerista bougainvillii*) lays eggs that take several weeks to hatch, whereas on Tasmania the young are born partially enclosed in a calcified shell, and have obviously almost completed their development within the mother because they hatch within a few days.

Live-bearing or ovoviviparous lizards include Jackson's chameleon (*Chameleo jacksonii*) and the flapjack chameleon (*Chameleo fuelleborni*) and several skinks, especially the larger species such as the Solomon Island skink (*Corucia zebrata*), the blue-tongued skink (*Tiliqua scincoides*), and the shingleback (*Tiliqua rugosus*). Other live-bearers are the tiny viviparous lizard (*Lacerta vivipara*), the slow worm (*Anguis fragilis*), the night lizards (*Xantusidae*) and the knob-scaled lizards (*Xenosauridae*).

The known cases of viviparity, in which the mother contributes nutrients for the development of the embryo, occur in some of the skinks, including the Brazilian species *Mabuya heathi* and the Italian three-toed skink (*Chalcides chalcides*). The young rely almost entirely on nourishment from the mother as the eggs are tiny and could not possibly support growth of the embryo to the point of birth. But this form of embryonic development has not been widely studied and is believed to be more widespread.

■ SOME OF THE SPECIES

Solomon Island Skink (*Corucia zebrata*)

This is the largest species of skink, a totally arboreal animal and the only one with a prehensile tail that is so important for its lifestyle it cannot be shed like other lizards. It is restricted to the Solomon Islands, and reaches a length of 26 inches (65 cm), and is olive green with dark vertical stripes on its back, with smooth and shiny overlapping scales, and sharp toenails for climbing. It has a large head and powerful jaws with sharp teeth, yet it is a folivore that eats mainly leaves and flowers plus soft

Broad-headed Skink   *Few lizards care for their eggs, and the skinks are the most protective species. Several, like the broad-headed skink of the southeastern United States, brood and guard their eggs, actually curling around or over them, and attacking other skinks or salamanders that try to steal them.*
Photo: Courtesy NPS, photo by Brent Steury

fruit, which is also unusual for a skink as the others are insectivorous. Furthermore, it is unusual in providing a high degree of parental care, the only live-bearing lizard known to care for its young, which is also rare in lizards that lay eggs, being known or suspected in a few other species of skinks. It is ovoviviparous (in which the egg hatches just before or when it is laid), and it has only a single offspring, born after an internal incubation period of almost seven months. The huge baby, which is almost half the size of its mother, is protected by her and occasionally also by the father, who attack anything threatening their baby, and apparently they also protect other baby skinks that are not their own.

Western Skink (*Eumeces skiltonius*)

The western skink is a small, shiny scaled lizard with a pointed head. It reaches a maximum length of 6 inches (15 cm) and has a broad, brown stripe down its back that is bordered by brownish-white stripes, beginning on the head and ending at the base of its brown tail. Young skinks have a brilliant cobalt-blue tail, but this fades and assumes their parent's coloration as they mature. It lives in a wide range of habitat, including rocky areas near streams, grassland, woodland, forest clearings and dry hillsides, across western North America west of the Rockies, from southern British Columbia to Baja California. It is diurnal and hides at night under rocks and logs, appearing at sunup to search for invertebrates such as beetles, crickets, spiders, and woodlice. In midsummer the female western skink makes a nest—a depression in loose, moist soil beneath a rock or log—and lays up to six eggs. She then stays close to them, usually curled around them, and protects them from nest robbers, mostly other skinks. She has been observed repeatedly basking and then returning to "cuddle" them, which has been interpreted as possibly raising her temperature to warm the eggs. Otherwise the ambient temperature is sufficient for their incubation.

Five-lined Skink (*Eumeces fasciatus*)

A native lizard of eastern North America, from southern Ontario west to Minnesota and south to Florida and Texas, this skink prefers forest edges, and cleared areas with rocky outcrops, logs, and tree stumps, with plenty of opportunities for sunbathing. A slender lizard with small limbs, it reaches a maximum length of 8 inches (20 cm) and has five creamy-yellow stripes on a brown background running along the back and sides from the snout to the base of the tail. Like the western skink, juveniles have a bright blue tail that fades with age. This skink lays many more eggs, however; up to 18, in a cavity cleared of debris beneath a log or rock, or even within an abandoned rodent hole, usually in an area where the soil has a high moisture content. Their eggs have thin parchment-like shells and absorb water from the soil. The incubation period varies from four to seven weeks depending on the temperature, and during this time the female broods them and defends them against small predators, usually other skinks. She curls around her eggs when the soil is moist, and rests over them to prevent moisture loss when the soil is dry. Brooding females have also been seen to periodically bask in the sun to raise their temperature and then return to the eggs. They eat rotten eggs and roll any that have been dispersed back into the nest with their snouts. Females may lay in communal nests and share the duties, making sure one stays behind to guard the eggs at all times. The hatchlings leave the nest the day after they hatch, and the parental care ends then.

■ SNAKES

Snakes are the second most varied group of reptiles with almost 3,000 species. They belong to the suborder *Serpentes,* and with their supposed ancestors the lizards form the order *Squamata.* The traditional belief is that they evolved from ancient lizards long ago, and lost their legs when they became adapted for burrowing in the soil. They are wide-ranging animals, occurring on all continents except Antarctica, and extend north into the Arctic Circle in Scandinavia, and south almost to the cold tip of South America. Yet they are absent from islands on which lizards still occur, including Ireland and New Zealand.

They are the most recognizable reptiles, with their smooth and elongated bodies, forked tongues, and staring lidless eyes. Their bodies are supported by up to 300 vertebrae and associated ribs that extend around the body and are attached to the ventral scales, allowing their fast sinuous movement as they literally pull their body over the scales by muscular action. They can swallow very large prey because the upper jaw is not fused to the skull but attached by elastic ligaments, and is divided to allow the two halves to move independently. This elasticity allows the snake to open its mouth very wide and then, using alternate movements, "walks" its jaws over the prey. As they lack a breastbone, their ribs can expand to accommodate the huge meal. Although there are legless lizards that resemble snakes, all snakes are legless; but they do have a tail, which is the section of their body beyond the cloaca. They have highly sophisticated senses of smell, taste, and heat-sensing.

Like the lizards, snakes reproduce by laying eggs or giving birth to living young. The egg-laying or oviparous species include the pythons, the grass snake, king snakes, milk snakes, rat snakes, and corn snakes. The live-bearing species may be ovoviviparous or viviparous, depending on the amount of internal care provided for the young. The former group, in which the embryo is nourished only by the egg yolk, include all the boas, vipers, rattlesnakes, and water snakes. Viviparous species, whose young receive nourishment from the mother during their growth, appear to be poorly represented in the snakes, although this may be because knowledge of the connection between mother snakes and their embryos is rather limited. However, at least the garter snake (*Thamnophis sirtalis*) is known to feed her embryonic young, as a placenta-like connection between the mother and her embryos delivers nutrients and removes wastes. The red pipesnake (*Anilius scytale*) is also truly viviparous.

Egg-laying is considered the primitive and ancestral form of reproduction in snakes; live-bearing developed when eggs were at greater risk, and natural selection then favored their retention by the mother and the birth of living young. The danger may have been from the climate, especially when living at high latitudes or elevations. Two examples illustrating this are the garter snake, which occurs on the Canadian prairies where winters are very long and very cold, and the adder or viper (*Vipera berus*), that lives within the Arctic Circle in Eurasia.

Snakes seem unlikely subjects to provide parental care, but some oviparous species brood their eggs, and others guard them; however, most bury or hide them. Even the aquatic oviparous sea snakes (some are live-bearers) come ashore to lay their eggs in rocky cavities or in the warm sand. The land snakes bury their eggs in loose soil or sand, in mounds of decomposing leaf litter, or in farm manure heaps or piles of wood shavings; they also hide them beneath rocks or rotting logs. Virtually any location where they will stay moist, and the site will be warmed by the sun or by decomposition, is suitable. Snakes do not leave their eggs exposed, as they would be unlikely to survive. The only time snakes' eggs are visible is when a python, or perhaps a bushmaster, is coiled around them, protecting and incubating them. However, no snakes care for their young. The egg-hiders abandon their eggs as soon as they have laid them, and none of the incubators care for their hatchlings, as the benefit of incubating lies only in caring for their vulnerable eggs.

The king cobras build a large nest of leaves and humus (they are the only snakes to make a nest) in which the eggs are laid—and then both parents guard the nest until the eggs hatch, with the female (or queen) coiled on top of the heap, and the male nearby. This is the only known example of a male snake providing parental care. Live-bearing snakes do not provide care for their young, which are totally independent as soon as they are born. In fact some snakes, and not just the cannibalistic species, regard their young as prey. Even captive boas (*Epicrates*) have eaten their young, which must have a place to hide immediately after they are born. Despite the unique care the king cobras provide for their eggs, they are cannibalistic to their young, and it has been suggested they may even leave the nest site just before the eggs hatch to avoid the temptation of eating their babies.

The pythons show the ultimate in parental care. All species really "incubate" their eggs, and it is a proven fact that many of them provide warmth generated through muscle contractions (shivering), which raises the temperature of the eggs. They are the only cold-blooded vertebrate[1] animals known to generate warmth in this manner. Their eggs adhere in a compact mass, allowing the female to coil around them and retain them for the whole incubation period (which may last for 8 to 10 weeks), during which time she does not feed and loses considerable body mass. However, this is inconsequential compared to the value of improving her reproductive success by protecting the eggs from predation, and preventing their dehydration and the desiccation of the egg yolk. The actual amount of energy spent on incubating eggs depends on the ambient temperature, and in warm and humid regions the female does not need to shiver to raise her temperature or that of the eggs. Brooding pythons make rhythmic muscular contractions every few seconds, raising their body temperature several degrees above the ambient temperature.

Whereas raising their body temperature to warm their eggs appears to be the prerogative of the pythons, other snakes do guard their eggs. Although most pit vipers are ovoviviparous (give birth to living young), the bushmaster (*Lachesis muta*), the Malay pit viper (*Calloselasma rhodostoma*), and some Asian pit vipers (*Trimeresurus*) lay eggs and guard them. This includes coiling around them, but it is unknown if muscle contractions or an increase in temperature is involved. In the southeastern United States, the large, aquatic mud snake (*Farancia abacura*), which feeds almost exclusively on amphiumas, lays its eggs beneath a log or even in an abandoned alligator's nest and then coils around them until they hatch.

■ SOME OF THE SPECIES

Indian Python (*Python m. molurus*)

This is one of the largest constrictors, reaching a length of 21 feet (6.4 m) and weighing up to 200 pounds (91 kg). It is a native of India, Pakistan, Sri Lanka, and Nepal, where it lives in a wide range of habitat, including river valleys, grassland, open woodland, rain forest, and swamps, preferring an environment with a permanent water supply. Its prey consists mainly of small mammals, such as monkeys, rodents, piglets, young deer and antelope, chevrotains, or mouse deer; plus lizards and birds such as peafowl and junglefowl. It is mainly an ambush predator, waiting patiently for its prey to pass within striking distance. It is a very attractive snake, heavily patterned with brown and golden-brown rectangles on a black background, but it is lighter and more golden in overall color than the other subspecies, the Burmese python (*P. m. bivittatus*).

Like the other pythons, which are all egg-layers, the female Indian python guards and broods its eggs. A solitary animal, she joins with a male only for mating, and about eight weeks later lays about 50 eggs. These stick together in a clump, and she then coils around them and rarely leaves, except perhaps to drink, during their 10-week incubation period. Every two or three seconds, she makes rhythmic muscle contractions, known as "shivering." Recordings taken during brooding, by inserting a probe between the snake's coils, show that her temperature can be raised 12.6°F (7°C) above the environmental temperature, although the increase is normally 5.4°F to 7.2°F (3°C to 4°C).

Bushmaster (*Lachesis muta*)

This is the largest New World snake, reaching a length of almost 12 feet (3.7 m). It is a pit viper (*Crotalidae*), a relative of the rattlesnakes, and therefore a member of the most specialized family of snakes adapted for hunting in complete darkness. This is achieved by a sophisticated heat-sensing system that commences in sensory facial pits, which can detect the heat emanating from warm-blooded prey and send a message to the snake's brain. The bushmaster lives in Central America and northern South America, plus Trinidad. It prefers tropical rain forest, but in Trinidad favors the cooler cloud forest. It eats small mammals, such as rats, agoutis, and opossums, and has long fangs that can inject a large amount of powerful hemotoxic venom, which affects the circulatory system of its prey and is fatal to humans without immediate medical attention. The bushmaster lays eggs, about 12 per clutch, often in an abandoned rodent burrow or in a hole beneath tree roots; it coils around them and defends them aggressively until they hatch, between 60 and 75 days later, depending on the environmental temperature.

King Cobra (*Ophiophagus hannah*)

The largest venomous snake, the king cobra grows to 16 feet (5 m) long and can rear up for one-third of its length, a very impressive and terrifying sight; but it prefers to avoid confrontations with people. The loose fold of skin on either side of its neck expands to form a cobra's "hood" when the animal is agitated, but it is unpatterned, unlike the other cobras. This giant but slender snake varies in color, and may be green, black, brown, or yellowish brown, but always has yellowish-white cross-bands. It lives in forest, grassland, cultivations, and mangrove swamps in Southeast Asia, Indonesia, and the Philippines. It has large, hollow, and erect fangs, and its venom is neurotoxic, affecting the respiratory centers in the medulla of the brain and compromising the transmission of nerve impulses to the muscles. It has a large amount of venom in keeping with its size. It is cannibalistic, its diet being mainly other snakes, both venomous and nonvenomous ones, that it easily overpowers.

The king cobra is a solitary snake that bonds just for breeding, the pair staying together from the time of mating until the eggs hatch. About two months after mating, the female (or queen) scoops up leaves with her body to make a nest pile, and then lays up to 50 eggs in the center, covering them with more leaves. She then coils on top of the pile and her partner stays nearby, guarding and patrolling the area against intruders. The eggs hatch after an incubation period of about 70 days, but both parents leave the nest just before they hatch, possibly to avoid the temptation of eating their own babies.

■ CROCODILES

The name "crocodile" actually belongs to a number of species within the reptilian order *Crocodilia,* which also contains the alligators, caiman, and gavials. So collectively they are really all crocodilians, but that name is generally only used by biologists. The crocodiles and birds, and the dinosaurs too, shared a common ancestor, the Archosaurs, that appeared on earth about 250 million years ago. About 20 million years later they diverged into two groups, the crocodiles in one, and the birds and dinosaurs in the other. For millions of years crocodiles shared the world with the dinosaurs, and the largest crocodiles were twice the size of the largest living species and are assumed to have preyed on herbivorous dinosaurs. They are considered primitive in the sense that they have changed little since the days of the dinosaurs, which signifies a very successful animal able to thrive for millions of years without needing to change its basic plan, and they continued to flourish after the dinosaurs became extinct. In fact, their survival was never seriously threatened until the arrival of modern man.

There are 23 species of crocodiles, separated into three families. The *Crocodylidae* contains the "true crocodiles," giants like the Nile crocodile and the estuarine crocodile. The family *Alligatoridae* contains the alligators and caiman; and the long-snouted gavials belong to the *Gavialidae.* They are all powerful semiaquatic carnivorous reptiles, with two exceptions[2] restricted to the tropical and subtropical regions of the world, as their preferred temperature range is 86.0°F to 95.0°F (30°C to 35°C). They cannot withstand temperatures below 43°F (6°C) and suffer heat prostration if their temperature rises over 104°F (40°C). Immersion in water enables them to cool down, and they also lose heat through gaping. In times of drought they burrow into the moist substrate of the riverbed and estivate, and both species of alligators—which have the most northerly distribution of all the crocodiles—hibernate in burrows during the winter. At night, when the land temperature is cooler, they stay in the water, which has been warmed by the sun, and by morning, when it has cooled down, they come out to bask again. They are perfectly adapted for life in the water, with sense organs in their jaws that can detect minute vibrations and movements, and a special palatal valve that allows them to breathe while holding prey in their jaws and when floating with just the tip of their snout above water. Although mainly freshwater animals, the estuarine crocodile is a marine species and the American crocodile prefers salt water. These species have salt glands on the back of their tongues for secreting sodium chloride.

All crocodiles lay eggs with very thick membranes and heavily mineralized shells, with a calcium content of 40 percent and 2.8 percent magnesium, more than chickens' eggs contain. Parents usually eat the eggshells and membranes after their young have hatched, to benefit from their minerals. When bird's eggs are incubated, they must be turned several times daily to prevent the embryo sinking to one side and sticking to the membrane, which affects its hatchability or may deform the chick. The crocodile embryo, however, actually becomes attached to the eggshell and turning the egg can kill it. Also, unlike birds' eggs, an air space does not develop in the crocodile egg, and the embryo acquires all its oxygen through the shell. During the incubation period, bacteria in the nest attack the outer calcified layer of the shell, increasing its porosity and allowing the passage of oxygen, as well as making it easier for the shell to crack at hatching time. But hatchlings frequently need assistance to escape from the eggshell. Crocodile eggs are sensitive to environmental conditions. They cannot withstand submersion in water for more than a few hours, and a very moist nest medium encourages fungal infections. They cannot withstand high heat, and if the nest becomes too dry, the evaporation of moisture from the egg can desiccate the contents.

In birds and mammals, the sex of the embryo is determined at fertilization; but in the crocodiles, as in some other reptiles, the incubation temperature of the egg affects the embryo's sex, as they lack sex chromosomes, so it cannot be determined genetically. When the incubation temperature is 87.8°F to 89.6°F (31°C to 32°C), the hatchlings will be males, while at temperatures above and below this they will be females. The incubation length is also affected by temperature, being extended by the lower values.

Crocodiles show the most advanced parental care of all reptiles. Some species build a nest of piled vegetation, sometimes on a base of soil and pebbles, then lay their eggs in a depression in the center and cover them. Others dig a hole in the riverbank for their eggs, but these must be above high-water mark to avoid being flooded. After laying her eggs, the mother crocodile, and in at least eight species the father also, guard the nest and protect the eggs. The very obvious nature of a pile of vegetation on a river bank or in a marsh is a great attraction to predators. These vary according to the crocodile's distribution. In South America, nesting caiman must keep watch for marauding coatis, peccaries, raccoons, and teguexin

Saltwater Crocodiles *Crocodilians show the most highly evolved tender loving care of all the cold-blooded animals. Even the saltwater crocodiles, armored giants reaching a length of 23 feet (7 m) and weighing up to 1 ton (1,017 kg), carry their hatchlings in their mouth to water, and then guard them there.*
Photo: Courtesy Adam Britton

lizards. In Africa, the Nile crocodile must guard her nest against mongooses, hyenas, and monitor lizards; while in India, wild pigs and jackals are the major potential predators of gavial nests. The parents can do nothing about the fire ants that get into cracked shells and kill the hatchlings.

Baby crocodiles call out when they are about to hatch, and the parents respond to their calls and open up the nest, but if they do not all hatch in one day they will re-cover the remaining eggs. Hatchlings frequently have difficulty escaping from their eggshells and the parents then help them, by gently crushing the eggs in their mouths or pressing against them with their legs or body. They then carry the young in their mouths to the water, where they form a crèche so the parents can more easily watch over and protect them. Otherwise they are independent and catch their own food—tiny fish and insects—but at the first sign of danger they rush back to the parent. The young may be protected by their parents for up to one year, each recognizing the other by chemical means. They are vulnerable to large fish and eels, storks, and otters, and other crocodiles also eat them.

■ SOME OF THE SPECIES

Nile Crocodile (*Crocodylus niloticus*)

This is a very large crocodile, vying with the gharial and estuarine crocodile as the largest species and reaching a verified length of 19.5 feet (6 m). It is restricted to Africa, excluding the Kalahari Desert; although small remnant populations still exist in the Sahara, and it also occurs in Madagascar. The Nile crocodile does not build a nest mound but digs a hole in a sandy riverbank or lake shore, above the water line, in which to lay its eggs. It may use the same site for many years, and breeds early in the dry season so the eggs will have time to hatch before the rains flood the nest and kill the embryos. A large female may lay up to 100 eggs, which normally hatch in about 85 days, but may take 10 days longer in cooler regions such as in South Africa.

The eggs are usually laid in layers, each covered with sand. The top layer of eggs sometimes has just a shallow covering, although the depth of the nest and its covering normally allows for the intensity of the sun and the texture and moistness of the substrate. The female guards the nest, sometimes by lying right across it or by resting nearby in the shade of riverbank trees, and the male is usually in the vicinity also. Unfortunately, the presence of the female advertises the fact that there are eggs for the taking, which is of interest to a host of potential predators, including monitor lizards, storks, hyenas, mongooses, baboons, bush pigs, and even vultures. The lizards are most successful as they work in pairs—one distracting the guardian, even decoying her some distance away from the nest, while its companion raids the nest, and the first one then returns quickly to share the spoils. When a nest has been raided and some eggs have been left exposed, the parents make no effort to re-cover them, although they may continue to guard the nest.

The young grunt to draw attention when they are due to hatch, and both parents assist those that have difficulty escaping from the shell by rolling them around gently in their mouths and crushing the shell. They also eat infertile eggs and embryos that are dead in the shell after cracking them open to check on their condition. The mother carries her hatchings to the water, stopping to pick up any that hatched and left the nest themselves, when they attract her attention with calls and tail twitching. The hatchlings are 11 inches (28 cm) long and are very vulnerable to a wide range of predators, including large fish, turtles, storks, herons, raptors, mongooses, and genets. They remain together in a group and are protected in the water by their mother for several weeks, communicating by yelping when separated, and submerging in response to her warnings of danger—which she does by vibrating her sides. They may sleep on her back or head. The babies eat small fish, water beetles and other aquatic invertebrates, and frogs and their tadpoles.

Gharial or Gavial (*Gavialis gangeticus*)

This crocodile's name is derived from *ghariyal,* the Hindi word for crocodile. It is one of the largest species, with recorded confirmed lengths of up to 19.5 feet (6 m). Its most unmistakable feature is its long and narrow snout, with many interlocking sharp teeth for catching fish, which form the bulk of its diet, and it lies in wait and catches them with a sideways snap of the jaws. It has attacked humans, but generally releases them and is not considered a man-eater. The gharial is a crocodile of deep and fast-flowing rivers in southern Asia, where high sandbanks provide nesting sites. There are now two separate populations—a western one in the lower reaches of the River Indus and other rivers in central western India, and an eastern one that has a much larger range. It occurs in the Ganges and Mahanadi rivers of India, the Brahmaputra and other rivers in Bangladesh and Assam, and the Irrawaddy River in Myanmar.

The gharial's nest is a hole excavated in a sandy or gravelly riverbank in March or April, and a female often digs several before being satisfied with the right combination of porosity and moisture of the substrate. She lays about 40 eggs, which need a temperature of 89.6°F to 93.2°F (32°C to 34°C) for their incubation. Both parents remain near the nest, to protect their eggs from sloth bears, jackals, striped hyenas, and monitor lizards. When the young begin to call, usually from within the eggs, the parents open up the nest and help them out of their eggshells. The mother carries them to the water, and they remain together until the monsoons begin and the rivers swell, which usually start a few weeks after they have hatched. Poor hatchability of the eggs, and the low survival rate of the young from predation and being washed away by the floods, reduces their survival rate to adulthood to no more than 1 percent of the babies.

American Alligator (*Alligator mississippiensis*)

The official state reptile of Florida, the American alligator is distributed throughout the southern United States from North Carolina south to Florida and then westward to Texas and Oklahoma. It is the northernmost-ranging crocodilian, its range reaching about one degree farther north than the Chinese alligator. It prefers freshwater swamps and marshes, and also rivers and lakes; it only occasionally enters brackish water, as it lacks the salt-secreting buccal glands of the American crocodile. In Florida it has reached a record length of 17.5 feet (5.3 m), but the average size of an adult male is 11 feet (3.3 m). The American alligator first breeds at the age of 10 years, and the female makes a mound of vegetation in June or July. Opening up the center of the mound, she lays up to 50 eggs and then covers them. The fermenting vegetation provides warmth for the eggs and prevents the nest from cooling down at night as it would if it was dependent only on the substrate.

Alligators lack the chromosomes that determine sex, so the sex of the embryos is dependent upon their incubation temperature, and the situation of the nest affects its internal temperature. Females are produced at temperatures below 86.0°F (30°C) and males at 93.2°F (34°C) or above. On dry levees, nest temperatures tend to be higher and produce more males, but not many alligators nest on the levees. In the marshes, the nest temperatures tend to be lower and produce more females. The incubation period is 65 days, and the mother stays nearby for the whole period to deter nest predators, of which raccoons are the worst offenders. The hatchlings make high-pitched grunting noises as they begin to break the shells; the mother responds and, with her feet and mouth, uncovers the eggs. Some have difficulty breaking out of the eggshell and she assists, cracking the shells in her mouth or with her body, and then carries them to water. The hatchlings remain in groups called pods or crèches for their first winter, usually in the vicinity of the nest, where she guards them until the following summer from their many potential predators. These include storks, raccoons, large fish, bobcats, and even other alligators.

Notes

1. Several flying insects can raise the temperature of their wing muscles by "shivering" prior to taking off.

2. The exceptions are the American alligator (*Alligator mississippiensis*) and the Chinese alligator (*Alligator sinensis*), whose ranges extend into the temperate regions, where they are forced to hibernate in winter.

4 Cold-blooded Chicks

The most recent evolutionary evidence supports the fact that modern birds evolved from the theropods, carnivorous dinosaurs with short forelimbs and strong hind legs that flourished from the Late Triassic Period (about 230 million years ago) until the close of the Cretaceous Period (65 million years ago). Fossils of small dinosaurs with feathers, discovered recently in China, prove that birds and dinosaurs shared a common ancestry and that the 10,000 species of living birds really are highly evolved dinosaurs.

During this long period of development, birds diverged to show a great variety of shapes, sizes, colors, and behavior. They learned how to survive in the tropics and in Antarctica. Some became flightless; others are so highly evolved for aerial life they never land on the ground. Their nesting behavior is also as varied, including where they nest, the number of eggs they lay, the length of their incubation, and their methods of chick-raising. But the major difference between the young of all birds, and therefore the care provided by their parents, is their degree of development and dependency when they hatch.

There are two broad categories of dependency in young birds. Most hatchlings are helpless, others are totally independent, and there is a range of conditions between the two. Chicks that are completely dependent on their parents or other birds for their raising are said to be altricial, whereas those that are more developed and are partially or even totally able to care for themselves have been categorized as precocial. The simple definition of the two categories is that precocial chicks can leave the nest as soon as they hatch, whereas altricial chicks are raised in the nest. Two other terms are also occasionally used to describe these categories, "nidifugous" for precocial chicks and "nidicolous" for mainly altricial chicks. Although it is the nestlings that are either altricial or precocial in their development, the terms are generally more widely used to denote the species rather than just their chicks. This chapter is concerned with the altricial species, the following one with precocial birds.

Both forms of embryonic development evolved as a result of food availability and the degree of pressure from predators, and the hatchlings differ in their development as a result of the type of egg from which they hatch. The egg that produces an altricial chick has a relatively small yolk, containing just over half the calories of the egg from which a precocial chick hatches. Consequently the altricial chick is less well developed, being naked, blind, and helpless, when it hatches. It is totally dependent upon its parents, who must immediately begin the task of finding food and keeping it warm. Altricial chicks have relatively smaller brains than precocial chicks when they hatch, but their rich food and more rapid growth result in their developing a larger brain than the precocial chicks by the time they are mature. An exception to this general rule are the parrots, which lay nutrient-rich eggs despite being altricial birds, and their chicks are consequently very intelligent.

Altricial chicks are certainly not cute and fuzzy "easter chicks." Initially they are uncoordinated and ungainly, with large heads and bulging eyes, distended stomachs, tiny wings, and large legs and feet. They are virtually helpless, only physically capable of holding their heads up and gaping for food. They are nest-bound and must be brooded by a parent as they are cold-blooded, for they cannot maintain their body temperature; their uncontrollable thermoregulatory system allows it to fluctuate with the environment. They cannot withstand low temperatures and must be protected from the elements, and would die without the warmth provided by their parents. Their thermoregulatory mechanisms do not begin to function for several days, and in the passerines may not be fully functional until they are well feathered when about ten days old. The emperor penguin's chick is brooded for 50 days before it can join a crèche of similar chicks, and the baby wandering albatross cannot generate its own heat until it is 30 days old. These chicks must also be fed regularly, and in most birds both parents are involved in their raising.

Altricial species include all the passerine birds (such as sparrows, chickadees, crows, flycatchers, and mockingbirds) plus hornbills, pigeons, pelicans, hummingbirds, vultures, and one gallinaceous bird—the hoatzin. They range in size from the world's smallest bird—the bee-hummingbird—to the great white pelican, one of the largest flying birds and the species with the largest altricial hatchlings. Altricial species mainly build their nests above ground, either in the form of an open nest in a tree, in which the eggs can be seen, or in holes in trees or sandbanks, where the young are out of sight until they fledge. But several altricial species, including the albatrosses and pelicans, nest on the ground, while some shearwaters nest underground. Conversely, some precocial species nest above ground (see Chapter 5).

There is also a subcategory of similar species that are known as semi-altricial birds. Their young have a covering of down feathers, but they are immobile and confined to the nest where they are fed and brooded, although their down provides some protection. Penguin chicks are considered semi-altricial, covered with fine down feathers, except the king penguin, which is naked when it hatches. In some semi-altricial species (the egrets, herons, and hawks) the chicks' eyes are open when they hatch, but in others (the owls) they open after a few days. But, whether they lack down upon hatching (and so are termed psilopaedic) or have down (and are called ptilopaedic), they are all helpless and unable to thermoregulate.

■ PARENTAL CARE

Parental care in birds is associated with their social and mating systems, and these vary widely as they have evolved to meet the challenges of survival. Practically all species are monogamous, the arrangement in which a male and female bond for the season or longer, sometimes for life. They are therefore able to share the duties of raising the young, on the basis that two parents can raise more chicks than a single parent; as the help of the male in raising the brood is known to almost double the number of young raised. Only 2 percent of birds are polygamous, which means the same as in human terms—having more than one mate at a time—which is then divided into polygyny and polyandry. Polygyny is the more common arrangement in which a male has more than one female partner and the females are left to care for the eggs and chicks. Polyandry, when a female mates with two or more males, lays eggs and leaves each one to incubate them and raise the young, occurs in less than 1 percent of birds. Excluding the African black coucal (*Centropus grilli*), the polyandrous species are precocial. A few birds are promiscuous, in which multiple individuals of both sexes are involved in breeding, and parental care is normally the female's duty.

It has puzzled scientists why male birds would become monogamous, as sharing the incubation and chick-raising duties is very confining and results in the loss of mating opportunities and therefore reproductive success. Surprisingly, parental care uses a lot of energy, because 90 percent of a bird's metabolism is directed to maintaining its body temperature, and incubating eggs—providing warmth for them through the brood patch—uses a lot of body heat. Female Canada geese (a precocial species) nesting on the Arctic tundra lose up to 40 percent of their body weight during their 28-day incubation period, due to heat loss and because they do not feed as regularly as normal. The investment in reproduction differs in the sexes, as it is a question of gamete production. Gametes are the specialized cells through which chromosomes are passed to the offspring—a small gamete or sperm from the male, a large gamete or egg from the female. A female's reproduction is limited to the number of eggs she can lay; a male's reproduction is limited to the number of mates he can possess. Therefore females gain more by being selective in their choice of mate, whereas males gain more by having several mates, and the low investment they must make in a polygynous mating arrangement favors sexual selection and evolution. When their eggs hatch, the monogamous parents' involvement requires an even greater investment in time and energy. However, it is now known that many monogamous birds practice "extra-pair matings" with others beyond their pair bond, and the young within a clutch can have different parentage.

Changes in hormone production, stimulated by environmental conditions, prompt reproduction. Prolactin, a hormone produced by the pituitary gland, is associated initially with reproduction and parental care in birds. Levels of the hormone rise during nest-building and egg-laying, peak during incubation, and then decline when the eggs hatch and other stimuli—visual and calling—stimulate the

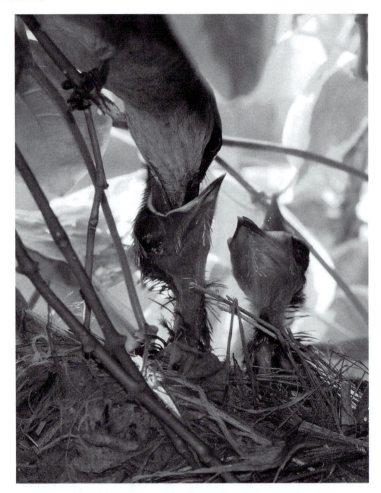

Yellow-vented Bulbuls *A typical altricial species, a yellow-vented bulbul brings berries and insects back to the nest to feed its naked and helpless young. Unable to thermoregulate, the chicks must be brooded at night until they are feathered. It is a common bird in Southeast Asia, Indonesia, and the Philippines, especially in cultivations and gardens.*
Photo: Loong Kok Wei, Shutterstock.com

feeding response. Higher levels of prolactin are known to occur in nonbreeding birds that were helping at the nest, and in the males that were involved in caring for their young. Most birds raise just one clutch of chicks annually, but some small passerines may have two or even three. Double clutching, as it is called, depends upon altitude and latitude. A species living in the lowlands may double clutch, whereas the same species at higher elevations of its range may not have time to do this. The short summers in the high north only allow birds to raise a single clutch of chicks.

With few exceptions birds breed annually, especially in the temperate north and south, where breeding is controlled by the seasons. Some birds leave a favorable climatic region to travel long distances and arrive at their breeding place before winter has ended. Food may be short then, but is available when their young hatch so they can complete their development in the brief summer. In tropical regions, breeding may occur at any time of year, although it is also governed by the climate, especially the arrival of the annual rains. The timing of the breeding cycle is determined by the potential availability of food, especially during the critical period—the actual raising of the young. Insectivorous species plan their nesting so that their chicks hatch at the start of the wet season, when insect larvae or locusts are hatching, to take advantage of the new plant growth. Fruit eaters wait until later in the wet season when fruit and berries are ripening. Seedeaters are last, when grass seeds are mature and beginning to dry at the end of the growing season. Birds that do not breed annually are the emperor penguin and some albatrosses, as they do not have time in one short summer to raise their young to independence.

All birds lay eggs, whose development cannot proceed without incubation, and finding a suitable site for them must be considered an aspect of parental care, even in the species that do not incubate their eggs or raise their young. Survival for the parasitic birds hinges upon finding the right nest at the right time, and sneaking in to lay their eggs in the host bird's absence. The megapodes, which bury their eggs where the sun or thermal activity incubates them, provide parental care in a very basic but essential form, whereas these birds are normally considered uncaring parents.[1] Consequently, all birds show parental care in varying ways, for not a single species can survive without the involvement of a caring parent at some stage after courtship and mating.

Parental care in birds is usually associated with nest-building, however, and most species build nests. These vary from the elaborate woven nests of the caciques and weavers, to the open nests of the crows, herons, storks, and eagles. Many nest in holes they have excavated themselves, drilled in trees by the woodpeckers and barbets, in sand banks by bee-eaters and sand martins, or even just on a flat sandy beach like the African and Magellan penguins. Others use natural cavities or usurp the tree holes of woodpeckers and barbets. The ovenbird and swallows build nests of mud, and with secretions of their own crops, the cave swiftlets make small cup-shaped nests high in dark caves. But some birds do not even make a nest and lay their eggs directly onto the ground, on a tree stump or on a cliff ledge. The emperor and king penguins incubate their eggs on their feet, and guillemots' eggs, laid on a bare rock ledge high above the sea, are conical in shape to prevent them from rolling off.

There are both benefits and risks associated with attending the nest during incubation and then brooding the chicks. Eggs are vulnerable, and the highest mortality in birds occurs during egg-laying and incubation. Parents provide warmth for their eggs and then food for their offspring; but protection in other ways, such as from predators, is less certain. Although nesting nonpredatory birds may successfully discourage a crow, a magpie, or even a hawk with their persistent mobbing, wherever altricial birds nest in the world, climbing predators are generally undeterred by a frantic parent. In many countries, domestic or feral cats are a major problem; in North America, gray squirrels and raccoons easily reach open arboreal nests. In South America, the tree-climbing tayra is a nest robber, while in southern Asia, civets, genets, and the binturong take eggs and nestlings. Most bird parents, and especially the passerines, are absolutely helpless against such nest robbers, and even the hole-nesting bee-eaters and barbets are not safe from snakes. Only the hornbills, sealed in their nest cavity, have almost complete protection from predation.

■ INCUBATION

Like the reptiles, in which the amniotic egg evolved, birds lay similar eggs, in which an amnion is present during their embryonic development. The altricial birds lay small eggs in relation to the hens' body mass, and they also have relatively smaller yolks than those of the precocial species. Consequently, there is less demand for energy during egg production, but a greater demand during the raising period. Incubation (and the eventual chick-raising) in the altricial species can be undertaken by the female or the male, or can be a joint venture of both parents, and in a few cooperative breeders several females may incubate. In the passerines the incubation is shared, with the hen doing most of the work; in the raptors it is always the female who cares for the brood while the male stands guard.

Sitting on the eggs does not necessarily denote incubation, however, for the mother's body warmth cannot reach the egg through her insulating feathers. Incubation, which initiates embryonic development, begins only when the parent exposes the brood patch on her abdomen to facilitate heat transfer to the eggs. Brood patch development is controlled by hormones that cause the feathers to fall out, and the fat beneath the skin to disperse, so there is no barrier between the birds' highly vascularized skin and the eggs. Brood patches occur in both sexes in species that share the incubation. Whereas practically all the birds that incubate their eggs do so with their brood patch, the boobies or gannets lack a brood patch and incubate with their feet. The eggs are held under the webs of their large feet, and as they near hatching they rest them on top of their feet.

There are certain "costs" to birds incubating their eggs, regarding the energy that is used and not replaced, especially for single parents. In most birds this cost is considered to be the amount of heat transferred from the bird's body via its brood patch that is sufficient for the development of the embryo. This is above the normal energy used daily by the bird in maintaining its own body temperature. Energy use will obviously be greater in cool temperate regions and especially in the case of species, such as the gray jay, that begin nesting while the northern forest is still snowbound. Energy use also varies according to the size of the eggs and the clutch. An incubating bird cannot leave its eggs for long during cool weather, and some birds, especially when they are the sole care-givers, rarely leave the nest at all, relying upon their reserves to sustain them. Broody chickens incubating pheasant eggs must sometimes be physically lifted off, and prevented from returning, before they will eat and drink. When birds are caring for their chicks, they seldom regain their condition as they are too busy feeding their chicks to eat properly.

In diurnal birds that share the incubation, the changeover between the parents must occur during daylight, so that each partner has an opportunity to find food, and in nocturnal species it occurs in reverse; but the changeover may not happen on a daily basis. Shearwaters may be away at sea for several days and then return to the colony under cover of darkness. Albatrosses may be absent for up to one week while their partner is incubating, and the male emperor penguin does not get relieved of his responsibility all winter, as the female does not return until spring. In some species the nonsitting bird is not actually off duty but must bring food back to its partner and then eventually to the chicks as well. In addition to keeping the eggs warm, the sitting bird must also turn them regularly to prevent the yolk dropping to one side and the embryo sticking to the membrane, and to ensure they are evenly warmed.

Birds' eggs need a specific temperature and humidity for their embryonic development, and these vary (a little) according to the species. Birds have a higher body temperature than mammals, averaging 105°F (40°C), and most species incubate their eggs in the range of 97°F (36°C) to 99°F (37.2°). Some, especially the penguins, have much lower incubation temperatures, averaging 96.5°F (36°C) and it is even lower in the emperor penguin, who balances the egg on his feet and covers it with the brood patch that is engorged with blood. This provides an incubation temperature of only 87.8°F (31°C), but the environmental temperature may be -76°F (-60°C).

Incubation periods are generally shorter for altricial species than precocial ones, as their chicks are not as highly developed when they hatch, but the periods are quite variable. Small passerines such as waxbills incubate their eggs for only about 11 days, whereas the largest passerine—the superb lyrebird (*Menura novaehollandiae*)—must incubate its single egg for 50 days. The emperor penguin holds its egg on its feet for about 65 days, and the wandering albatross incubates its single egg for about 80 days, sharing with the precocial kiwi the record for the longest egg-incubation period. Most birds wait until they have laid their whole clutch before beginning to incubate, so that their eggs hatch at the same time; but the birds of prey begin when they lay their first egg, so their eggs hatch with a gap of several days between them. This gives the older chick the advantage in species that practice siblicide. One exception is the pygmy owl, which lays its complete clutch of eggs before beginning their incubation.

■ RAISING

All altricial birds care for the chicks they have hatched, whether they have laid the eggs themselves or have incubated those of a parasitic species. Warmth and food are the immediate priorities for the naked and helpless young, and may be provided by the mother, father, or both parents. Warming their chicks, instead of eggs, is called brooding, and its length is determined by the chicks' ability to thermoregulate. That, in turn, depends not only on the species, but also on the weather, the location and type of nest, and the protection it affords the chicks. Nestling woodpeckers in a tree cavity, for example, would be more protected than baby crows in an open treetop nest. For the first two or three days, altricial nestlings must be brooded constantly, while the other parent provides the food. Brooding time is reduced as the chicks grow and produce feathers, and eventually is only needed at night, by which time they are generally too big for a parent to cover anyway. While this arrangement is generally true of altricial chicks, there are a few species, such as the swifts, whose nest-bound young can become torpid for several days in cold weather when food is short.

After sitting on their eggs for at least 11 days, or considerably longer by many species, the arrival of the chicks has a profound effect upon the parent birds, and the change from incubating eggs to feeding the chicks is prompted by their visual stimulation. After the relatively peaceful days of incubation, it is the beginning of a period of intense activity. In most passerines, both parents must make an amazing number of food trips daily, a great drain on their health. There have been many observations made of the number of trips birds make daily to the nest with food, with the record probably going to the pair of house wrens that made 1,200 visits to feed their large brood during one long summer day. In many species the male provides food, while the hen broods the chicks. Male raptors hunt and bring food back to the female, which she dismembers and feeds to the young. As they grow they are able to feed themselves, but if the hen is lost before they are old enough, they will starve alongside food left on the nest by the male. In the tree-nesting horn-bills, the male brings food to his partner sealed into the nest cavity. In the red-eyed vireo, the reverse occurs—the hen brings food to the male brooding the chicks.

The manner of feeding altricial young varies. Passerines carry insects in their bills and stuff them into their chicks' gaping mouths. The herons, storks, gulls, albatrosses, and shearwaters regurgitate food straight into their chick's beaks, and the hummingbird inserts its rapier-like bill deep into its chick's esophagus to pump in the nectar from its crop. In another version of this arrangement, pelican, cormo-rant, booby, and penguin chicks put their heads into the parent's beak and reach for the food in the crop. Great tits may squeeze the juices from large caterpillars into their chicks' beaks and then discard the skin. Prey is carried in the feet by the owls and hawks, and they initially tear it apart for their newly hatched young, and feed tiny pieces of bone also, because they appreciate the importance of calcium for their growing chicks. Young birds are generally given the types of foods their parents eat, and which they themselves will eat when they are independent. But there are excep-tions, and the most unusual of these is the production of "milk" by birds. Mammals are not the only animals to feed their young initially with body secretions or prod-ucts. Some frogs lay eggs to feed to their tadpoles, baby discus fish eat secretions from their mother's skin, and a number of birds feed their young with secretions of the crop or esophagus.

For the first three or four days of their chick's life, parent pigeons regurgitate highly nutritious epithelial cells sloughed off the walls of their crops. This "pigeon's milk" is rich in fat and protein, and both the male and female doves and pigeons feed it to their young irrespective of their adult diet—whether they eat seed or fruit—and so provide the protein that the parent's diet lacks. The parents then begin introducing solids to their young (called squabs), but continue feeding milk until they are fledged. Pigeons have only one or two squabs per clutch, and by producing milk for them they can raise them quickly and have several broods in a season.

Flamingos also produce milk for their chicks, which need a liquid solution for their first two months until they develop their specialized filtering beaks. Flamingo milk is very rich in fat but has less protein than pigeon's milk, and is produced by

Great Egrets *Great egret chicks have a covering of white down feathers when they hatch and their eyes are open, so they are considered semi-altricial chicks. But they are helpless and immobile, and need to be brooded to keep warm. Both parents feed the nestlings, which start to clamber around the neighboring branches when they are three weeks old.*
Photo: Lori Skelton, Shutterstock.com

glands lining the upper digestive tract. The milk is bright red due to the presence of the carotenoid pigment canthaxanthin, which maintains the pink coloration of flamingo feathers. Like the pigeons, in a highly unusual evolutionary development, both parents produce this milk for their chicks. The only other bird known to feed its young with a body secretion is the emperor penguin, in which the male incubates the single egg on his feet for two months during the Antarctic winter. If the female has not returned when the egg hatches, he feeds the chick for a short time with a secretion produced by the lining of his esophagus. On her return she regurgitates fish and squid for the chick, but many are lost when their mothers are delayed.

A bird's crop, an expanded section of the esophagus, has an important role in feeding the young, for in many species it is the organ in which food is stored temporarily for taking back to the chicks. Digestion does not occur in the crop as digestive enzymes are not secreted there, as they are in the stomach, but some breakdown of the food may commence. Most seed-eating birds feed insects to their young initially, but in the few that feed regurgitated seed (such as the goldfinch), the seed is softened in the crop before being fed. The hoatzin has the most highly developed and enlarged crop, in which bacteria ferment its leafy diet, and this leaf pulp is then regurgitated for the chicks, which are altricial despite being gallinaceous (chicken-like) birds. Vultures go a stage further, for the food they take back to the young in the nest enters their stomachs, where strong acids partially digest the decomposing meat, skin, and bones that form their diet, and presumably also destroy potentially harmful bacteria, before they regurgitate it for their chicks. When vulture chicks are hand-raised, as California condor chicks were for their conservation, their food must first be "predigested" with hydrochloric acid and digestive enzymes.

Most altricial nestlings beg for food by gaping, opening their beaks wide and reaching up as far as their scrawny necks allow, an action that triggers the feeding response in the parents. It is such a great stimulant that other birds of the same species, passing the nest on the way to feed their own chicks, have been unable to resist it and have stopped off to feed the strangers. Feeding the chicks of another species is less likely to happen, however, as the insides of many chicks' mouths are brightly colored and are specific to the species, and a mother will only feed her own young. The young of hole-nesting birds, such as kingfishers in which the nest is at the end of a long and narrow burrow, are stimulated by the change in light as the parent arrives at the entrance hole; and as they mature they take turns rushing to the entrance to greet her, as there may only be room there for one chick.

In addition to keeping their chicks warm and feeding them, parent birds must also keep the nest clean, and they are aided by the fact that the feces of most young have a gelatinous covering, allowing the sac to be picked up and carried away from the nest site to prevent attracting attention. Initially parents may stimulate their chicks to defecate by pecking gently at the anal region or pulling at the down feathers there. As the chicks mature, they eject their feces onto the edge of the nest and the parent removes them, except during the last days of raising when they may be left to build up around the edge of the nest. Swallow chicks differ, as they eject their feces over the edge of the nest onto the ground below, because the risk of a predator reaching them high under the eaves of a house or barn is very low. Parrots and other hole nesters do not remove the droppings, and they are dried by the rotting wood that forms the base of the nest; but in some species, such as the woodhoopoes, the chick's copious feces make the nest very messy. It is impossible, however, for some birds to hide their nesting activities. There is no doubt about the location of a heronry, from the noise of the birds above and the ground below streaked with their liquid droppings, typical of fish-eaters. But the danger to them comes more from flying predators.

Parents visit the nest quickly to feed their young, for the longer they are there, the greater is the attention drawn to it. They generally feed the strongest chicks first to be certain that some will survive if times are suddenly hard, and feed the weaker individuals only when the strongest are satisfied. If food becomes short, they may simply continue to feed the strongest young and neglect the weaker ones. If one of the parents of a monogamous pair is lost, the workload of the remaining parent may be too much for it to handle. Albatrosses, despite incubating their eggs for over two months, and then spending up to eight months feeding their chick, will abandon it if food becomes scarce and their own survival is jeopardized.

The risk of death in the nest does not come only from outside sources; some young birds are not safe from their nest mates. In the parasitic species, it is the alien chick that kills or ejects its nest mates. But there are also birds in which the danger comes from their own siblings—behavior that is called Cainism, fratricide, or siblicide. It occurs in boobies and skuas, but is most well known in the raptors, several of which raise only one nestling despite laying two eggs, but these hatch at intervals and the younger chick is immediately at a disadvantage. The older chick may be directly aggressive and harass or intimidate its younger sibling, which usually dies in its first week. This harsh behavior ensures one chick gets adequate food and stands a good chance of surviving. Fratricide occurs in several species of eagles, including the Spanish imperial eagle (*Aquila adalberti*), the lesser spotted eagle (*Aquila pomerina*), Verreaux's eagle (*Aquila verreauxi*), and in the related secretary bird (*Sagittarius serpentarius*). The bald eagle (*Haliaeetus leucocephalus*) may lay three eggs, but it is rare for all three chicks to survive. In the golden eagle (*Aquila chrysaetos*), Bonnelli's eagle (*Hieraaetus fasciatus*), and the African hawk eagle (*Hieraaetus spilogaster*), the second chick may survive.

■ HELPERS

In almost 300 species of birds, the parents have the help of others when raising their chicks. These "helpers at the nest" or "cooperative breeders" are generally close relatives, and this unusual nesting behavior occurs when breeding opportunities are limited due to the lack of suitable habitat, and not every female in the group can breed. There are several advantages to helping the parents raise their young. It places the helpers in a good position if anything happens to the parents, for they can then assume the territory. There are also genetic advantages if the helpers are related to the breeding pair, as it improves the chances of breeding and raising relatives. Young birds that help to raise siblings learn from the experience and then make better parents themselves, as first breeding attempts in birds that lack this experience often fail.

To improve the chances of reproductive success, cooperative breeding is practiced by the acorn woodpecker (*Melanerpes formicivorus*), Florida scrub jay (*Aphelocoma coerulescens*), groove-billed ani (*Crotophaga sulcirostris*), the green woodhoopoe (*Phoeniculus purpureus*), colies or mousebirds (*Colius*), caiques (*Pionites*), and conures (*Pyrrhura*). This unusual behavior involves several pairs making a single nest, in which all the females lay their eggs and then share the incubation and raising duties. This may seem altruistic, but in fact they are actually very selfish, at least in some species, as each female tosses out other females' eggs and may kill nestlings in an attempt to improve the survival chances of its own offspring. South America's guira cuckoo (*Guira guira*) lives communally in groups of up to 12 adults and the females all lay in one nest, laying perhaps 20 eggs in total. But then the co-operation ends, and the egg-tossing and chick-killing begins, as each female attempts to ensure its own genes are perpetuated.

Another survival technique involving bird helpers is the gathering together of chicks in a crèche, where they are supervised by a few adults, generally non-breeding birds, which allows the parents to gather food for their young without leaving them completely unattended. This is considered unusual behavior as the chicks are guarded by unrelated birds. Crèches of chicks occur in birds that breed in colonies and whose eggs hatch close together. It reduces the risk of predation to the individual—this is called the dilution principle—as a chick can lose itself in the masses and reduce its risk of predation. This does not reduce predation but does lessen the individual's odds, as the predator, a skua for example, is still going to get a baby penguin, but most likely one that is not in the crèche. Crèching there-fore improves the chances of a bird's genes surviving over those of birds that nest singly, thus favoring the practice through selection and evolution. If there is a greater survival rate of chicks in crèches, then more young from there will contrib-ute their genes to the next generation and thus favor crèche raising.

The colonial-nesting penguins (the emperor, king, Adelie, gentoo, and rock-hopper penguins) all put their young in crèches, with up to 100 chicks in each. When the parents return, they all recognize and feed their own chicks even in the largest colonies. Penguins join crèches when they have a good covering of down feathers, between three and four weeks old in the rockhopper penguin, and at about seven weeks in the emperor penguin. Pelican chicks also form crèches; the Australian white pelican's chicks gather together when they are three weeks old and stay there for two months, by which time they are fledged. There may be several hundred young pelicans in a crèche. The American white pelican is believed to be unable to totally thermoregulate until it is almost three weeks old, and soon after-ward leaves the nest to join other chicks in a crèche. The chicks of colonial nesting terns, such as the elegant tern, royal tern, and Sandwich tern, also join crèches; and when they are one month old, greater flamingo chicks leave their nests and join a crèche that may contain several hundred young birds that continue to be fed by their parents, although by then they can filter their own food.

Imprinting is most well known in the well-developed precocial species, when some youngsters become attached to the first moving object they see. It also occurs in altricial species but later, when their eyes open, and is especially well known to breeders of birds of prey. A hand-raised peregrine falcon, or a kestrel, may be so imprinted it will attempt to copulate with its human "partner's" hat; a practice that is actually of great value when sperm is required for artificial insemination pur-poses. Hand-raised California condor chicks must be fed with a hand-puppet like-ness of their parent's head while the keeper hides behind a screen, to avoid them becoming imprinted on humans.

American White Pelican Crèche *Pelican chicks need brooding by one of their parents initially, but when they can thermoregulate—maintain their own body temperature—which is believed to start just before they are three weeks old, they leave their nest and congregate with other chicks in a crèche supervised by several nonbreeding adult pelicans. This allows both parents to go fishing to meet their chicks' increasing demand for food.*
Photo: Courtesy NPS, photo by Bryan Harry

■ PARASITES

Many birds have totally surrendered their parental care duties, having successfully tricked others into undertaking them. Known as brood parasites, they lay their eggs in other birds' nests, and the birds that assume the tasks for them are called brood hosts. Although some of their eggs may be rejected, they have invested less energy in them and lay eggs in many nests. Birds who delegate the rearing of their chicks to another species do not bond, and after being mated the female goes off on her own in search of a suitable nest.

About 1 percent of all birds are parasites, who rely totally upon others to undertake all the parental duties. They do not build nests, incubate their eggs, or raise the young, but merely lay their fertile eggs in other birds' nests. They are members of five families, of which four produce altricial hatchlings: They are the cuckoos (*Cuculidae*), honey guides (*Indicatoridae*), blackbirds (*Icteridae*), and weaver finches (*Ploceidae*). They are called obligate brood parasites, as they must find a host bird or they cannot reproduce. The fifth family is the *Anatidae*, of which only the black-headed duck is a parasite, but a precocial one. Others, which may either impose on other birds or incubate their own eggs, are called facultative brood parasites, but they are also precocial species. A few parasitic birds are intraspecific, as they lay their eggs in the nests of their own kind; the cliff swallow (*Petrochelidon pyrrhonota*) is one of these, but they are mostly interspecific, as they use the nests and care provided by unrelated species. The cuckoos are the most famous brood parasites, in which 50 species in the Old World and three in the New World (out of a world total of 143 species) lay their eggs in other bird's nests.

The major advantage of practicing brood parasitism is that birds can lay more eggs. They are not restricted by the number of eggs they can cover and incubate, and they save energy in the long term as they neither incubate their eggs nor raise the chicks. The European cuckoo may lay 24 eggs in a season, one per nest, when it usually removes one of the host's eggs. But the hatchling cuckoo ensures it has no competition in the nest, for during its first few days it pushes all its companions (eggs and chicks) out, wriggling beneath them and heaving them over the edge. Whydah nestlings are more tolerant and grow alongside the host waxbill's young. The most intriguing aspect of the European cuckoo's parasitism is that it selects host species that are very much smaller, and the foster parents raise a single cuckoo chick that eventually fills their nest and is many times larger than its care-givers. Some parasitic birds have learned to mimic the host's egg coloration and markings, and the specific markings inside the nestling's bills to which their mothers respond. Yet the eggs of others are totally unlike those of the host, and consequently are often discarded, but the large number of eggs they lay compensates for this.

Africa is home to most parasitic birds. A group of seed-eating species called whydahs or widow-birds have evolved to take advantage of other birds, their hosts being small finches known as waxbills and firefinches, and unlike the northern cuckoo they are mainly host-specific. The paradise whydah (*Vidua paradisaea*) lays its eggs in melba finch (*Pytilia melba*) nests, and the pintailed whydah (*Vidua macroura*) in nests of the common or St. Helena waxbill (*Estrilda astrild*). The thick-billed cuckoo (*Pachycoccx audeberti*) parasitizes the nests of the plum-colored starling (*Cinnyricinclus leucogaster*), and the great spotted cuckoo (*Clamator glandarius*) is a large bird that lays its eggs in the nests of the pied crow (*Corvus albus*), but its nestling does not heave out the host's eggs or chicks and is raised with them.

Southern Asia is also rich in parasitic cuckoos. The large hawk cuckoo (*Hierococcyx varius*) concentrates its attention on different host species, and lays two kinds of eggs to suit both occasions. It can lay a bright pale blue egg to mimic those of the laughing thrushes (*Garrulax*) and the blue whistling thrush (*Myophonas caeruleus*), or brown eggs similar to those of the streaked spider hunter (*Aracnothera magna*). The red-winged crested cuckoo (*Clamator coromandus*) lets laughing thrushes and ground thrushes (*Zoothera*) incubate its eggs and raise the offspring, while the koel (*Eudynamys scolopacea*) lays its eggs in the nests of house crows (*Corvus splendens*) and, in atypical parasitic behavior, may lay several eggs in one nest.

In the New World, the giant cowbird (*Molothrus oryzivora*) parasitizes the nests of caciques (*Cacicus*) and oropendolas (*Psarocoleus*). Botflies (*Philornis*) are also a major parasite of these birds, laying their eggs on the chicks. As the cowbird chicks preen the cacique chicks and remove the botfly larvae from them, the presence of adult cowbirds around the colony is tolerated. However, they have a competitor in the form of a social wasp (*Polybia*), whose presence deters the botflies, and when the caciques have wasp protection they drive the cowbirds away—one protector is enough. The North American yellow-billed cuckoo (*Coccyzus americanus*) and black-billed cuckoo (*Coccyzus erythropthalmus*) occasionally lay their eggs in the nests of other females of their own kind, and of other species, but more often they build their own nests and raise their chicks, so parasitism is not mandatory and they are therefore considered facultative brood parasites. The brown-headed cowbird (*Molothrus ater*) lays only after the host has laid two or more eggs, and she pierces the shell of a host egg and carries it off to eat, replacing it with one of her own. Although her eggs are often not a good match and are discarded by the host female, the cowbird is a very successful brood parasite, to the extent that the populations of several host species are beginning to suffer.

■ TAKING OFF

Birds cannot expend energy on raising their young to the extent of jeopardizing their own lives, and they contribute sufficiently to giving their chicks a reasonable start in life without lowering their own chances of survival and future reproduction to the danger level. The length of care varies according to the species, but is determined by certain factors. Generally, the longer the period of care, the better the chances of the offspring's survival; but offsetting the length of the caring period is the fact that shortened care may allow the raising of a second brood, and the energy used during the extended care may affect the birds' ability to migrate or survive the winter.

All activity must be fueled, and the reproductive process places a strain on the bird beyond its normal living requirements and the obtaining of sufficient calories for its own needs. So there are certainly savings in energy costs and risks to the parents if their young fledge and fly early. When the risk of predation is high, even a day or two less in the nest can reduce the odds considerably. As nests are such vulnerable places, it is also in the chick's interest to leave as soon as possible. They are encouraged to do this by their parent's behavior, when they decrease the rate of feeding or call to them from outside the nest to entice them to leave. Fledging in altricial birds—when the chicks have attained their flight feathers—coincides in most species with their leaving the nest. But even after they have fledged and flown, the young of many altricial species continue to beg for food. Others, like the semi-altricial owl chicks, leave the nest long before they have fledged and perch on branches nearby for perhaps two weeks while their parents continue to feed them until they can fly. The pygmy owl is an exception, for its fledglings fly when they leave the nest, and as it started its incubation only after laying all its eggs, also unique in owls, its young are all the same size.

For most passerine or perching birds, the raising period (from hatching to fledging) averages three weeks, but several altricial species spend much longer in the nest. The larger species of albatrosses, such as the wandering albatross (*Diomedea exulans*), are cared for in the nest by their parents for nine months. The California condor has an even longer period of parental care of the young. They take eight months to fledge, and then receive post-fledging care for another six months. Fortunately, the male shares the raising duties, bringing much of the food to the young.

The termination of care by parent birds generally tapers off gradually, but in some species it is very abrupt. Penguin chicks are abandoned by their parents, and must swim and fish without assistance. Oil bird chicks, having grown very fat on their diet of palm nuts but not yet in possession of their flight feathers, are abandoned by their parents and survive on their stored fat until they can fly. Sooty shearwaters, whose chicks are collected commercially as muttonbirds, abandon them before they can fly. The chicks grow rapidly and are soon double the size of their parents; but before they are fully fledged, the parents go to sea to molt. The down-covered chicks live off their fat stores for three to four weeks after they have been abandoned and lose weight rapidly, but while doing so they grow their own flight feathers. When these are fully grown, they also take off for their long migration north to Alaska, where they spend the northern summer.

■ SOME OF THE SPECIES
Wren (*Troglodytes troglodytes*)

A tiny, plump bird, just 3.5 inches (9 cm) long, the wren has a short upright tail, and a rufous-brown body barred with darker brown and pale underparts. It lives in Europe and northern Asia, from where the most northerly populations go south for the winter. It resides all year in the British Isles, where it occupies a wide range of habitat, including suburban gardens, hedgerows, and woodland. During very cold weather, 20 or more birds may roost communally, nestled together in a garden nest box. The wren usually nests in a hole or crevice or in thick ivy against a wall, building a large domed nest of leaves, grasses, and moss with a small side entrance and an interior lined with feathers. Its breeding season begins in April, and it may produce two broods each summer. It is a typical monogamous passerine or perching bird, in which both parents share the egg-incubation and chick-raising duties. The usual clutch is six eggs but occasionally several more are laid, which places a great strain on the parents when they hatch after 15 days of incubation. The chicks are naked, blind and helpless—other than being able to gape for food. The wren is a highly insectivorous species, and the whole day is spent finding spiders, caterpillars, aphids, crane flies, and other insects for its nestlings. Hundreds of collecting trips must be made daily to satisfy their hunger, but their growth rate is very quick and they fledge when they are two weeks old.

Mourning Dove (*Zenaida macroura*)

The mourning dove is the common wild pigeon of the New World with a distribution from southern Canada south into Central America and the Caribbean islands. Birds that summer in the northern parts of this range move farther south for the winter. It is a slender, brown bird, with a bluish crown, a pink-tinged breast, and long central tail feathers; the outer white-tipped ones are visible in flight. Mainly a seed-eater, frequenting suburbia, fields, and woodland, it is the most plentiful North American pigeon since the demise of the passenger pigeon. The mourning dove is a prolific breeder that normally produces three broods yearly in the warmer parts of its range. Its clutch is just two eggs, however, laid in a flimsy nest of twigs in a tree or tall shrub, and both parents share the incubation period of 15 days and the raising of the squabs. Both parents feed them for their first three or four days solely with the sloughed-off walls of their crops, which contain nutritious epithelial cells rich in fat and protein. This "pigeons milk" is fed by all doves and pigeons to their young, irrespective of their adult diet. Mourning dove parents then begin introducing seeds to their squab's diet but continue feeding the milk until the young are fledged. This occurs when they are just two weeks old, but the parents continue to feed them for at least another week.

Emperor Penguin (*Aptenodytes forsteri*)

The largest living penguin, and the largest flightless waterbird, the emperor penguin weighs 65 pounds (30 kg) and is 45 inches (1.1 m) long. Its body is pale bluish gray with a black border extending from the sides of its neck to the flanks, with white underparts that have a yellowish tinge on the upper breast. Its head is blackish blue with large yellow and white auricular patches, and its long bill is also dark with a thin red line along the bottom mandible from the gape almost to the tip. Its legs and feet are dark gray. It is confined to the Antarctic continent, where it lives in the waters immediately adjacent to the land mass, and rarely strays outside the Antarctic Circle.

The emperor penguin is mainly a squid-eater; it may dive down to 750 feet (228 m) in its search for food and can stay submerged for 18 minutes. Despite the harshness of the climate it breeds in midwinter, unlike all other species, because the summers are too short for its lengthy courtship, egg-laying, incubation, and chick-raising duties, which require at least five months to complete. As winter approaches at the end of March, the penguins come onto the ice shelf to breed. They nest mainly on the coast, but many also make long treks inland, often up to 200 miles (320 km) to Cape Crozier on the Ross Ice Shelf, tobogganing over the ice on their bellies. Courtship, egg-laying, and incubation all take place in total darkness during the long winter night, which lasts four months. After laying their eggs, the females pass them over to their mates, who each balance an egg on their feet, and must keep it covered with a flap of their belly skin to maintain a temperature of 87.8°F (31°C). The females return to the sea, and the males alone incubate the eggs, fasting and eating snow for moisture, and in the process losing almost 50 percent of their body weight. To conserve heat, their metabolic rate drops and they become sluggish and huddle together, packed 10 birds to a square yard (.83 square meters), changing places regularly so that each gets a turn in the middle, where they are protected from the blizzards. The incubation period is about 65 days, and the males feed the chicks with an oily crop secretion until the females return with their crops full of squid. They then take over and the males make the long journey back to the coast. When the chicks are about 50 days old and well covered with thick brown down, they join a crèche, where they are chaperoned by a few adults while the others go off to feed.

Chinstrap Penguin (*Pygoscelis antarctica*)

Since the decline of the baleen or whalebone whales and the greater availability of shrimp-like krill, the chinstrap penguin has increased in recent years and is second to the macaroni penguin as the most abundant species of the northern Antarctic Ocean. It now has an estimated population of about 7.5 million breeding pairs, with enormous rookeries on the shores of the Antarctic Peninsula and on islands in the Scotia Sea—South Georgia, South Sandwich and South Shetland. The largest colony, on Zavodovski Island in the South Sandwich group, comprises 10 million birds, and there are smaller rookeries on Heard, Kerguelen, and Bouvet islands. The chinstrap penguin is a wanderer that has strayed as far as Tasmania and Mac-Quarie Island, and it is also probably the most nocturnal of the penguins, as it feeds at night as well as during the day. It returns to its nesting colonies in November, and the juveniles are ready to depart the rookeries in March. It has a faster breeding rate than the crested penguins of the genus *Eudyptes,* which show preferential treatment to the strongest chick and never raise more than one. Like them, the chinstrap penguin lays two eggs in late November, but hatches and raises both chicks, showing no favoritism and feeding each one equally. Both parents share the 35-day

Chinstrap Penguins   *A penguin of the northern Antarctic, the chinstrap penguin nests in enormous colonies on the Antarctic Peninsula and on islands in the Scotia Sea. It generally raises two chicks, unlike several related species that lay two eggs but normally only raise the strongest chick. There is no weaning period for these chicks; they fledge when two months old and take off to sea to cope on their own.*
Photo: Courtesy U.S. NOAA

incubation duties—in shifts of seven or eight days—and the chicks spend 30 days in the nest before joining a crèche. They fledge and go to sea when they are 60 days old. The breeding results are poor, however, when sea ice persists close to the nest colony and the parents cannot get to the open sea to feed. Starting in March, they leave the breeding colonies and move north of the pack ice for the winter. The main predators of this species are leopard seals at sea, and skuas on land.

European Cuckoo (*Cuculus c. canorus*)

This bird is a summer visitor to the British Isles and Europe, and spends the winters in Africa. It is a well-known bird of the British countryside, its "cuckoo" call unmistakable. It is also the most familiar of the birds that parasitize others, allowing them to incubate its eggs and raise its chicks. Without the bother of caring for its own, the cuckoo can lay lots of eggs, up to 24 in a summer season, allowing for the rejections that potential hosts do make. It does not carry its egg in its beak from elsewhere as previously thought, but lays it directly into the host bird's nest, generally just one per nest, removing one of the host's eggs at the same time. When more than one cuckoo's egg is found in a nest it is usually the work of different birds.

There is great variety in the size and shape of cuckoos' eggs. Some closely mimic the host's eggs in color and size, but in other nests are totally unlike them, and sometimes even smaller despite the cuckoo's much larger body size. At 14 inches (36 cm) long, it is usually much larger than its brood hosts, such as the hedge sparrow (*Prunella modularis*) and sedge warbler (*Acrocephalus shoenobaenus*), both little more than 5 inches (13 cm) long, and the wren (*Troglodytes troglodytes*) is even smaller. The hen cuckoo must deposit her egg in the host's nest at exactly the right time, no easy task when birds defend their nests once laying starts and rarely leave them for long, so there is a very short window of opportunity. The cuckoo must be mated, have an egg ready to lay, find a suitable nest, and then wait for its opportunity when the occupant leaves. She can apparently retain an egg in her oviduct for up to two days awaiting the opportunity to lay, and then does so very quickly, usually in less than one minute. As the cuckoo's egg-incubation period is shorter than that of the hosts, her egg hatches first. The naked nestling has a sensitive depression in its back, and anything touching this stimulates a pushing reflex, so it gets the host's eggs onto its back, one at a time, and heaves them out of the nest. The cuckoo chick is believed to simulate the calls of a whole clutch of its host's chicks, stimulating the foster parents to feed it unceasingly, creating the incredible sight of a single chick filling the nest and a parent bird actually perched on its back feeding it.

Great Horned Owl (*Bubo virginianus*)

The only eagle owl of the New World, this large species has a wide range from the Arctic to the tip of South America, where it lives mainly in woodland but also in farmland and rocky arid canyons. It is sooty brown mottled with grayish brown above, with darker cross-barred underparts and a contrasting white collar. But it is geographically variable, with dark birds in the northern forests and pale brown ones in arid regions. It has prominent ear tufts and an orange-gray facial disc, weighs up to 4 pounds (1.8 kg), and stands 24 inches (61 cm) tall. Like most species of owls, the female is slightly larger. The great horned owl is a very powerful bird, easily able to catch rabbits, squirrels, rats, ducks, grouse, quail, and occasionally domestic cats and poultry. It can overcome newborn white-tailed deer fawns and is the only owl to regularly eat skunks.

The great horned owl nests early in the year, but does not build its own nest. Instead it either usurps those of hawks, herons, crows, or squirrels, or it nests in a large tree hollow or in a cave, and it is very aggressive in defense of its nest. Its clutch usually contains four eggs that are incubated for 30 days while the male brings the food. The nestlings are semi-altricial, helpless, and blind, but with a covering of fine down feathers, and their eyes open a few days after hatching. Eggs are laid daily, but incubation begins with the laying of the first egg, so they hatch at the same frequency, and there is always a size difference between the chicks. They leave the nest when six weeks old, before they are feathered, and perch on nearby branches for another three weeks until they fledge.

Great Indian Hornbill (*Buceros bicornis*)

The largest hornbill, about 4.5 feet (1.4 m) long and with a wingspan of 5 feet (1.5 m), the great Indian hornbill is a mainly black bird with a band of white feathers on its wings and a banded black-and-white tail. Its neck is yellow, it has a black face with a naked red eye ring, and a huge yellow bill topped with a large casque, made of keratin, that is not as heavy as it looks. This hornbill is a native of the tropical rain forest from India through Southeast Asia and also Sumatra, where it supplements a mainly fruit diet with insects, birds' eggs and chicks, and possibly small lizards and tree frogs.

Like all the tree-nesting hornbills, it is a cavity-nester; but it is not equipped like the woodpeckers to make its own hole, so it seeks a natural cavity high in a large rainforest tree, with a suitable entrance hole large enough to enter but small enough to be sealed almost closed. During the courtship that follows, the male brings the female fruit and continually checks the cavity to assure her of his dedication. If she accepts him and the nest site, they mate and she then enters the cavity, and with a mixture of regurgitated fruit, feces, and mud that sets very hard, the entrance hole is closed except for a narrow slit, and the female is enclosed. She then lays, incubates, and raises her young while entombed in the nest cavity, during which time the male faithfully feeds her and the young. The egg incubation period is approximately 40 days, and it takes another 80 days for the chicks to fledge; so the female is confined for four months, until the seal is broken, but she has raised her chicks in relative safety from predators.

Groove-billed Ani (*Crotophaga sulcirostris*)

A slender, all-black bird about 13 inches (34 cm) long, including its long tail, the groove-billed ani is a member of the cuckoo family *Cuculidae*. But unlike so many cuckoos, it does not parasitize other birds. Its most distinctive feature is the very large beak, with parallel grooves along its length and a curved upper bill. It is a Central American and northern South American species that just reaches into the United States in southern Texas. It is mainly insectivorous, but also eats small lizards, snails, and occasionally fruit. Groove-billed anis are cooperative breeders that live in small groups comprising several monogamous breeding pairs, with one or two unpaired helpers, that defend permanent territories. All the members of the group help to build a single communal nest, in which the females of each pair lay their eggs and then take turns with the incubation duties and raising the chicks, aided by the helpers. However, they are not exactly as cooperative as this type of arrangement appears; for with several females laying in a nest and only one bird incubating, there is a problem with surplus eggs. Like the primary female ostrich who ejects excess eggs (laid by other females) from her mate's nest but is careful not to toss out her own, each female ani attempts to eject the eggs of the other layers. But there are often still too many eggs in a nest to incubate them all, and eggs become buried beneath the top layer and do not hatch, so they can never really be sure that their own eggs have hatched anyway.

Green Woodhoopoe (*Phoeniculus purpureus*)

This bird lives in wooded regions and in the riverine forests of the savannahs across sub-Saharan Africa, excluding the rain forest zone. It is highly insectivorous and finds its food in tree bark and on the ground. Despite its name, it is more purple in color than green, has a very long tail, and reaches a total length of about 17 inches (42 cm). Adults have a long and curved thin red bill, and therefore live up to their alternate name of red-billed woodhoopoe. It is a communal species, living in small groups of up to 12 individuals, comprising a breeding pair and their helpers, all of which cooperate to defend their territory. Within each group, however, only one pair breeds, laying up to four blue eggs in a tree cavity, and all the other members of the group then assist in raising the young, bringing insects for them, feeding them, and brooding them. They roost communally, to conserve energy on cool nights. There appears to be little advantage to the other females in merely assisting the breeding pair in this way, other than the experience they gain in helping at the nest will assist them if they ever have the opportunity to nest themselves.

Notes

1. Although megapodes are also called mound builders, they do not all build mounds of soil and vegetation and allow the heat of decomposition to incubate their eggs. The eggs of some species are incubated by geothermal heat or by radiation.

5 Warm-blooded Chicks

Unlike the naked and helpless cold-blooded chicks that cannot thermoregulate, the newly hatched young of many birds are well developed. They have down feathers, are mobile, and their eyes are open, so they can leave the nest soon after hatching. They can respond quickly to changes in the ambient conditions and maintain their own body temperature, although most need brooding during cold and wet weather and at night. They are the warm-blooded or precocial chicks, the cute and fuzzy "easter chicks," more advanced when they hatch because the eggs that produce them have almost double the calories of altricial birds' eggs and require a longer incubation period. To produce such rich eggs, their mothers must receive sufficient food prior to egg-laying.

As they have a higher degree of independency, precocial chicks initially have relatively larger brains than altricial chicks, but their brains do not keep pace with their slower body growth and at maturity are smaller than those of adult altricial species, due to the much richer food they received as babies. They are less vulnerable because they can leave the nest soon after they hatch, and then scatter, hide, or freeze when threatened. However, there is a cost to this behavior: as they are still individually vulnerable to predation, their parents must provide protection, which may expose them to danger. Precocial birds range in size from the Chinese painted quail (*Coturnix chinensis*), just 4.5 inches (11 cm) long, and the slightly larger Inaccessible Island rail (*Atlantisea rogersi*), the world's smallest flightless bird; to the wild turkey (*Meleagris galloparvo*)—the largest flying bird—and the ostrich (*Struthio camelus*), the largest bird of all. They include all the waterfowl and gallinaceous birds (excluding the hoatzin), the ratites, rails, shorebirds, grebes, loons, bustards, plovers, trumpeters, cranes, and the megapodes.

There are several degrees of precociality. The megapodes are considered superprecocial species. They find a suitable site for their eggs and let the sun, decomposing vegetation, or volcanic heat hatch them; and their chicks are completely independent, find their own food, and can fly short distances when they emerge from the ground. All the other precocial species, except the parasitic black-headed duck, depend on their parents for brooding to keep them warm and dry, for protection from predators, and in most cases for assistance in getting their food. There is also a group of birds (the gulls and terns) whose chicks are considered semi-precocial. Their eyes are open and they have a covering of down when they hatch, but although they are partially mobile, they stay in the nest and are fed by their parents. This is obviously a good tactic in large colonies, where chicks could get lost or be attacked by other parents, for infanticide is known to occur in some gull colonies.

The methods of feeding precocial chicks also vary, from very attentive care to none at all, with the chicks finding their own food. The waterfowl, shorebirds, and ratites (except the quite independent kiwi chick) follow their parents, but find their own food; the chicks of the pheasants, partridge, jungle fowl, and other game birds also follow their parents but are shown what to eat; whereas young rails, grebes, loons, and cranes are initially fed by their parents. The members of one family of birds, the sandgrouse (*Pteroclidae*) of the deserts of Africa and Eurasia actually provide their chicks with water. The male flies several miles to the nearest water hole, soaks his belly feathers, which can absorb up to .70 ounces (20 ml) of water, and returns to the nest where the chicks pull on the feathers to retrieve the moisture.

Precociality in birds has its limitations, however. It is most suitable for species that nest on the ground (although several do nest in trees), and that take small foods from the surface of the ground or water. It is unsuitable for birds, such as the swifts, that hawk insects aerially and then return to an elevated nest, or for those that capture large prey and return to their chicks with a carcass. Although the longer incubation period of their large-yolked eggs means that the chicks are more developed when they hatch and require less care, it also results in the incubating bird spending more time on the ground incubating its eggs or sheltering its young, where it is vulnerable to predators. However, the ability of the chicks to scatter and freeze when danger threatens lowers the risk of all the chicks succumbing to an attack, which is usually the outcome when a predator finds a nest of immobile altricial chicks. Precocial chicks rarely stay in the vicinity of their nest.

■ PARENTAL CARE

Like the altricial brood parasites that rely on other birds to raise their young, a number of precocial species also behave in the same manner. The black-headed duck (*Heteronetta atricapilla*) is an interspecific obligate parasite, as it cannot survive without the help of other birds. It lays its eggs in coot's nests, and never attempts to nest and incubate its own eggs. There are also many other examples of parasitism in the waterfowl. Some are interspecific parasites like the redhead (*Aythya americana*) that parasitizes mallard (*Anas platyrhynchos*) and lesser scaup (*Aythya affinis*) nests, but also nests and raises its own young, so it is called a facultative brood parasite as it is not totally dependent upon parasitism. The common goldeneye (*Bucephala clangula*) is similar, combining parasitism of the nests of other cavity-nesting ducks with her own nesting efforts. The Canada goose (*Branta canadensis*), snow goose (*Chen caerulescens*), canvas back (*Aythya vallisneria*), shoveller (*Spatula clypeata*), and several other waterfowl all lay eggs in conspecifics' nests as well as nesting and raising their own young. The ring-necked pheasant (*Phasianus colchicus*) is also known to lay eggs in other pheasants' nests. As the chicks of these species are precocial and self-feeding, they have less impact on the host than the parasites of altricial birds, where chick-raising is an energy-consuming business. Some waterfowl also practice what is called "dump nesting," when more than one female lays in a nest that may eventually have 60 eggs, far too many to be incubated. Wood ducks (*Aix sponsa*), muscovy ducks (*Cairina moschata*), and the hooded merganser (*Mergus cucullatus*) are known to do this, and it is believed to be connected to a shortage of nest sites.

The megapodes neither incubate their eggs nor raise their chicks, and they vary in their nesting behavior. Some build large mounds of soil and vegetation and rely on the heat of decomposition to incubate their eggs, but the males control the temperature of the mound, so they are considered caring birds. Others lay their eggs in soil or sand heated by radiation or by geothermal activity, and then leave them, so they are generally considered non-carers. However, site selection for the incubation

Canada Goose *Canada goose goslings are precocial and mobile. They have a coat of soft down-feathers and are immediately able to find food (generally grazing grass) without assistance from their caring parents. But they must shelter at night and when it rains, while their parents do their best to keep them dry and warm.*

Photo: Bruce MacQueen, Shutterstock.com

of their eggs is a most important task and deserves to be considered an aspect of parental care, similar to the reptiles that bury their eggs to protect them and to provide the right conditions for their incubation. To compensate for the lack of care after laying, megapode eggs contain the extra energy needed for their longer incubation period and greater embryonic development. They are much larger than those of the other gallinaceous birds, and may weigh between one-fifth and one-quarter of the bird's body weight (similar to the kiwi's egg). Their yolks may comprise 60 percent of the total egg weight, compared to the domestic hen's 33 percent. This must be considered parental care.

Mating systems vary considerably in the precocial birds, being associated with the demands for parental care. For example, if both parents are needed to raise the clutch, then they are monogamous, with the father and mother sharing the incubation and the raising duties. Monogamy features widely in the precocial species. Waterfowl are mostly monogamous; the swans, geese, and whistling ducks forming long-lasting pair bonds, whereas most ducks mate just for the season. Like many monogamous altricial birds, the mallard (*Anas platyrhynchos*) and the gadwall (*Anas strepera*) form pair bonds for the season, but may also mate with others, which is called extra pair copulation (EPC), resulting in extra pair paternity (EPP). Research has shown that matings outside the bonded pair occur frequently in many monogamous birds, so that the chicks in a nest may have different parentage.

Many gallinaceous birds are also monogamous, including the wild turkey (*Meleagris gallopavo*), the eared pheasants (*Crossoptilon*) and argus pheasants (*Argusianus*), plus the Hungarian or gray partridge (*Perdix perdix*), the California quail (*Callipepla californica*), and the northern bobwhite quail (*Colinus virginianus*). Most megapodes are monogamous, exceptions being the brush turkey (*Alectura lathami*) and the wattled brush turkey (*Aepypodius arfakianus*) that are promiscuous; the males make and maintain the nest mounds, which the females visit to be mated and to lay their eggs. The familiar ring-necked pheasant, an alien but now North America's premier game bird, is a harem species, the male mating several females within his territory, and then helping to guard the nests and the broods. In the precocial lek species such as the greater prairie chicken (*Tympanuchus cupido*) and the sage grouse (*Centrocercus urophasianus*), several males congregate and display to attract females, which are mated and then leave to incubate their eggs and raise their young without help.

Most ratites—the rheas (*Rheidae*), emus, and cassowaries (*Casuariidae*)—plus their ancestors the tinamous (*Tinamidae*) all practice the form of polygamy known as polyandry, in which the females mate with several males and lay eggs, which the males then incubate and raise the chicks. The other ratites—the ostrich (*Struthionidae*) and the kiwi (*Apterygidae*), share the parental duties. However, the kiwi chick is virtually independent upon hatching, and receives very little care from its parents, although its legs are weak initially and it does not stand until it is four days old.

The cranes (*Gruiformes*) are monogamous and form strong pair bonds that generally last for life, and their relatives the rails and the semiaquatic moorhens and coots are also monogamous. The shorebirds (*Charadriiformes*) have a variety of mating systems. Most plovers and lapwings are monogamous; and the sandgrouse (*Pteroclidae*) are also monogamous and nest solitarily, although close enough to give the impression of being colonial nesters. There are also many examples of polyandry in the shorebirds (with just one exception polyandrous birds are all precocial[1]). The African jacana (*Actophilornis africanus*) lays four eggs and then leaves the male to incubate them and raise the chicks. The spotted sandpiper (*Actitis macularia*), dotterel (*Chardrius morinellus*) and the red-necked phalarope (*Phalaropus lobatus*) are also polyandrous, the females mating with several males and then leaving each to cope with the incubation and brooding. The normal coloration of birds is reversed in these species, with the males being duller to camouflage them while they are incubating. The pectoral sandpiper (*Caladris melanotus*) is highly unusual as it is promiscuous; males mate with multiple females and females mate with many males.

Most precocial birds are ground-nesters, and their nesting arrangements vary from no nest at all to a huge pile of vegetation. The megapodes are also misleadingly called mound-builders, although several species do not build mounds but rely on volcanically heated soil or a sun-warmed beach to incubate their eggs. These non-mound-building species include Pritchard's megapode (*Megapodius pritchardi*), whose eggs are incubated by geothermal heat; and the maleo (*Macrocephalon maleo*), which lays its eggs in the sand above high tide on the beaches of Sulawesi, where they are hatched by solar heat. In these species the female digs a deep hole, lays an egg and covers it, then returns at intervals of several days to lay other eggs. Megapode chicks hatch after an incubation period of about 70 days compared to the domestic chicken's 21 days, and they may take two days to dig their way to the surface. After this long development period they are well feathered and can fly short distances, and have sufficient egg yolk left to support them for several days. The megapodes that build mounds and rely upon the microbial decomposition of vegetable matter to produce the warmth to develop their eggs, include the Australian species—the brush turkey (*Alecthura lathami*), orange-footed scrub fowl (*Megapodius reinwardt*), and the mallee fowl (*Leipoa ocellata*). The males scratch soil and vegetation into a large mound in which the hens lay their eggs, and with their very sensitive beaks, or possibly their tongues, they read the temperature of the mound and then regulate it throughout the incubation period by removing or adding soil above the eggs. None of these birds provide care for their chicks, which are totally independent upon hatching, but their mothers have provided extra yolk in their eggs to compensate for this lack of care, so they are well provided for.

Cranes make a nest of mud and reeds, while the swans build an enormous mound of vegetation for their eggs. The moorhen's nest is made mainly of the leaves of reeds and flags, with small twigs and often pieces of paper or plastic garbage, usually in the vegetation on the bank of a pond or stream; but occasionally they use an old crow's nest, or build their nest in full view on the semi-submerged branches of a tree. The grebes bring water plants up from the pond bottom and make a large, soggy pile just above the water line. Most gulls nest on the ground, making bulky nests of seaweed, heather, and feathers, but this depends entirely upon the materials on hand. The ground-nesting ducks make more elaborate nests of grasses, usually lined with down feathers from the female's breast. Many ground-nesting precocial species do not make a nest, but like several plovers, stone curlews, and the semi-precocial terns, they merely make a rudimentary scrape in the sand or pebbles on a beach and lay their eggs directly onto the substrate. Plovers often line their nests with tiny pieces of shell or small white pebbles. The nests of the phalaropes and sandpipers are a similar rudimentary scrape, but may be scantily lined with grass.

Surprisingly, many precocial species nest above ground, and their flightless nestlings must get down to ground-level soon after they hatch, as the parents do not bring food to the nest. Several precocial species in tropical America nest in trees. The currasows and guans (*Cracidae*) are tree-nesters, as are the trumpeters, the sun bittern and the limpkin, whose chicks must free-fall to the ground or water soon after they hatch. But they are outnumbered by the tree-nesting waterfowl, which includes the numerous members of the tribe *Cairinini*—the perching or wood ducks—such as the Carolina wood duck (*Aix sponsa*), the mandarin duck (*Aix galericulata*), maned goose (*Chenonetta jubata*) and the muscovy duck (*Cairina moschata*). Other tree-nesting ducks include the hooded merganser (*Mergus cucullata*), common merganser (*Mergus merganser*), the common goldeneye, bufflehead (*Bucephala albeola*), and the black-bellied whistling duck (*Dendrocygna autumnalis*). These birds all nest in tree cavities, using either old pileated woodpecker holes or natural cavities formed by age or lighting. They do not begin to incubate their eggs until they have laid their whole clutch, because the ducklings must leave the nest together and are enticed out by their mother calling from below.

As in the altricial species, the parent's warmth cannot reach her eggs through her layer of insulating feathers, so incubation only begins when the abdominal "brood patch" is exposed to facilitate heat transfer to the eggs. The incubation periods of precocial species are generally longer than those of the altricial birds due to the larger egg yolk and the chicks' greater degree of development upon hatching. It ranges from just 16 days for the Japanese quail (*Coturnix japonica*) and Chinese painted quail (*Coturnix chinensis*) and rises to the megapode's average of 70 days and the kiwi's egg that takes about 80 days to hatch. In proportion to the female's size, the kiwi egg is the largest, weighing 25 percent of the hen's body weight, whereas the ostrich's three-pound (1.3 kg) egg is only 1.5 percent of her body weight. The well-developed nature of the precocial birds, especially their eyesight, has resulted in the phenomenon of the chicks imprinting upon literally the first moving object they see upon hatching. The famous zoologist Konrad Lorenz will be forever associated with the graylag goslings that followed him everywhere.

An unusual feature of some birds is the ability of their chicks to hatch synchronously despite the eggs being laid at intervals. This occurs in the grassland ratites such as the ostrich and emu. The male does not feed his chicks, so they must leave the nest soon after they hatch; they cannot wait several days until all the eggs have hatched. The most advanced chicks therefore call in the eggs, and the less developed embryos then accelerate their development, permitting hatching synchronization and allowing the male to leave the nest with the majority of his potential brood.

Precocial chicks leave the nest soon after they hatch and must be prepared to avoid predators. Land species can run almost immediately, but a far better strategy is to avoid drawing attention to themselves. In response to their parent's alarm calls, they flatten to the ground and rely on their cryptic coloring. Shorebird chicks crouch when alarmed and their finely spotted down feathers break up their outline, and they blend with the sand, pebbles, and vegetation. Megapode chicks can fly as

Domestic Chicken   *Just like its ancestor the junglefowl, a domestic chicken draws the attention of its newly hatched, precocial chicks to a source of food. It does not feed them, but initially they must be shown what to eat, which is typical of the gallinaceous birds. To train them to eat by example, breeders usually put a larger, established chick with day-old pheasants and partridges hatched in incubators.*
Photo: Todd Evans, Shutterstock.com

soon as they appear above ground, and young curassows can become airborne within a few days. Aquatic species, such as the coots and moorhens, can swim under water soon after they hatch. Grebe chicks can also swim and dive, but rarely enter the water for the first few days and are carried on the parent's backs. Ducklings and goslings follow their parents into water immediately after they leave the nest, or as soon as they can get to the nearest pond; and the tree-nesting ducks usually nest over water so their ducklings can drop straight in. Swans also carry their cygnets on their backs while they are young, and the male jacana carries his chicks beneath his wings, pressed against his body. The sun grebe takes chick-carrying to its ultimate, actually placing them in pouches on its sides. Weka rails are the only birds known to pick up their chicks in their beaks and carry them away from danger.

Most ground dwellers, even the flightless ostrich and rhea, employ the "broken-wing" technique, letting one wing hang loosely as if broken, to draw predators away from their nest or chicks. The western sandpiper (*Calidris mauri*) and the rock sandpiper (*Calidris ptilocnemis*) use the "rodent run," running fast from the nest in a crouched position, resembling a rat and attracting the predator's attention. Birds that lay their eggs in the same place for generations are particularly vulnerable.

On the remote South Pacific island of Niuafo'ou, feral cats regularly visited the thermal nesting sites of Pritchard's megapode along the caldera's rim, and began waiting by the mist nets spread over the nest sites, effectively ending my attempts to catch the hatching chicks for a zoo conservation project.

In total contrast to the concern shown for their young, several birds leave their chicks completely in the care of others soon after they hatch. In unusual behavior that combines aspects of helping and crèching, they let others raise their chicks for them. Female king eider ducks in Iceland abandon their nests when their eggs hatch and leave their ducklings under the care of other females who gather them into groups and act as chaperones, shepherding them to the water to feed. European shelducks (*Tadorna tadorna*) have evolved similar behavior for when they fly to traditional molting sites in the North Sea from the British Isles, leaving their ducklings in the care of nonbreeding young adult ducks that have retained their wing feathers.

Despite their warm-bloodedness and their covering of down feathers, most precocial chicks must be brooded[2] by their parents during wet and cold weather and at night, and many must find their own food. The independent megapode chicks cannot even follow their parent's example, but many other precocious species also find food without assistance. Cygnets and goslings begin to graze soon after they hatch, and day-old mallard ducklings scoot over the pond surface chasing insects or finding pond weed. Others are shown what to eat by their parents. The parents of young gallinaceous birds—pheasants, partridges and domestic poultry—peck at suitable foods and even pick up items and drop them in front of their chicks; pheasant breeders do the same to entice day-old chicks to eat. The cranes and rails (including moorhens and coots) carry food to their chicks and offer it to them in their beaks. Precocial chicks grow fast, their growth rate being one of the fastest of all the vertebrates. Well-fed ostrich chicks can gain 52 ounces (1.5 kg) weekly; Japanese quail are sexually mature at just six weeks old; and the red jungle fowl, ancestor of all domestic poultry, breeds at five months of age. Artificial selection and genetic manipulation have produced broiler chickens large enough for the commercial market when just seven weeks old.

Fratricide or siblicide, which frequently occurs in some altricial birds, is less common in precocial species. Crane chicks are the worst offenders, being so aggressive to their siblings that both chicks in a clutch are seldom raised. Conservation programs have benefited from this behavior by collecting an egg from each wild nest and artificially raising the chicks, effectively doubling the reproductive rate of each breeding pair. But when this is done, the chicks must be penned individually to avoid conflict, while allowing them to see their neighboring siblings so they are not imprinted on humans.

■ SOME OF THE SPECIES

Elegant Crested Tinamou (*Eudromia elegans*)

A ground-dwelling bird resembling a small, compact chicken, with a slender neck and small head, the elegant crested tinamou is about 16 inches (40 cm) long, and weighs 1 pound (454 g). Its plumage is dull gray with darker streaks, white on the face and neck, and a long crest that tilts backwards with the feathers then turned forward at the end. It has strong, well-developed feet with just three toes, and no hind toe. The crested tinamou is a bird of southern South America, where it lives in the arid and semiarid grasslands, open woodland, and thorn scrub of Argentina and Chile, from sea level up to 8,200 feet (2,500 m) on the Andean slopes. It is an omnivorous species, eating seeds and vegetation in winter and mainly invertebrates—insects, worms and mollusks—in summer. The tinamous are believed to be the ancestors of the ratites, yet they did not become fully flightless, although they are poor fliers. Tinamous are promiscuous; a male mates with several females, each of which lays an egg in his nest and then goes off to find other male's nests to lay in. The males are known for their faithful care of the eggs, which may number up to 12, incubating them for 20 days and rarely leaving them to feed. Unaided, they then raise the highly precocial chicks, which have cryptic, dull-colored down feathers.

Brush Turkey (*Alecthura lathami*)

A chicken-like bird, about 28 inches (70 cm) long, with an all-black body, a bare red head, and a yellow throat wattle, the brush turkey is the largest of the three Australian megapodes, all of which build nesting mounds. It has sturdy legs and large toes, and scratches in the leaf litter for insects and fallen fruit in the rain forests and scrub bush of eastern Australia, from Cape York south to Sydney. The brush turkey's breeding behavior is one of the most remarkable of all birds. The male scratches together a huge pile of soil and vegetable matter, which may measure 13 feet (4 m) in diameter and stand 39 inches (1 m) high. Several females are attracted to the mound and, after being mated, lay their eggs in its center. Up to 50 eggs have been found in one mound. The females then seek another nesting male and repeat the process. So both sexes have more than one mate and are considered promiscuous (or polygnandrous), in which there is no lasting association between the sexes. The eggs are incubated by the heat of decomposition, which is controlled by the male. He maintains a temperature of 95°F to 100.4°F (35°C to 38°C) by digging out the center of the mound, inserting his highly sensitive beak to gauge the temperature, and then removing or adding soil as required. He also "protects" his mound, but is helpless against marauding goannas (monitor lizards), feral pigs, and dingos, other than repairing the damage to his nest afterward. The brush turkey's incubation period is about 50 days, much shorter than most megapodes, but when the chicks dig their way out they are fully feathered and independent, and can fly short distances immediately.

Black-headed Duck (*Heteronetta atricapilla*)

This duck is an intraspecific brood parasite, a bird that lays its eggs in others' nests. It prefers red-gartered coots (*Fulica armillata*) and red-fronted coots (*F. rufifrons*) to undertake the care of its eggs and ducklings, but may also lay in nests of the rosybill (*Netta peposaca*). As it makes no attempt to care for its own eggs or ducklings, and its survival is therefore totally dependent upon others, it is called an obligate parasite. The black-headed duck is a member of the stiff-tailed (*Oxyur-ini*) group of diving ducks, but rarely dives, usually upending like the dabbling ducks to feed on the pond bottoms. It lives in northern Argentina and neighboring Chile and Paraguay, and it is a vegetarian, preferring a diet of seeds. There is no lasting pair bond in this species; after mating, the female goes off to find a suitable nest, and has been seen forcing an incubating rosybill duck off its nest so she could lay her own egg, but she does not remove or destroy the host's eggs.

The black-headed duck is believed to lay only one egg in each host's nest, but they are apparently often rejected as they do not resemble the host bird's eggs. She must find a nest in which the incubation has not begun, although the embryonic duckling may be able to synchronize its own hatching to coincide with the other eggs. The incubation period is 24 days, and the duckling has a thick, downy coat and is completely independent, immediately going off on its own and not requiring brooding by its foster parent. It is therefore the only super-precocial (totally independent) species other than the megapodes, and is considered a benign parasite as it has no harmful effect on the eggs or young of its host, unlike many brood parasites.

Emu (*Dromaius novaehollandiae*)

The emu reaches a height of 6 feet (1.8 m) to the top of its head and 48 inches (1.2 m) to its back, and is the second-tallest bird after the ostrich. Weighing 85 pounds (35.5 kg) when adult, it is not quite as heavy as the cassowary. It is a dark, grayish-brown bird with a sparsely feathered, bluish-gray neck, and has long and powerful legs and thick toes characteristic of the ratites. It lives across Australia, west of the Great Dividing Range, but avoids the tropical forests of the north. It is a common bird, although the emus of Tasmania, Kangaroo, and King islands have been extinct for many years. The female emu usually has just one mate, but she may occasionally be polyandrous and have primary and secondary mates, and after leaving the first male to incubate the clutch of eggs she has laid, she then actively seeks another unpaired male who will then incubate her second clutch. After laying her eggs she has no further role in the process of reproduction. Emus nest between April and November, in the Austral winter. The male makes a flat bed of leaves and grasses, usually beneath a tree or large shrub, and placed to give a clear view of the surrounding country. His temporary mate then lays up to 15 dark grayish-green eggs, initially at four-day intervals, then every other day. The male is believed to commence incubation after the first egg is laid, but the hen continues to lay eggs beside him, and he pulls them under his body also, without leaving the nest. For the whole 56-day incubation period, he rarely leaves the nest, neither to eat nor drink, and can lose up to 22 pounds (10 kg). He turns the eggs up to four times daily and aggressively defends them. Cheeping noises can usually be heard through the eggshell after 50 days, and this stimulates some degree of hatching synchronization, despite the great interval between the first- and last-laid eggs. The chicks are fully precocious, clad in striped black and tan down, and are cared for by their father for at least six months although they find their own food, mainly insects initially.

Emus *The female emu lays her eggs in her mate's nest and then lets him incubate the eggs and raise the chicks. He sits solidly for the 56-day incubation period, rarely leaving the eggs even to drink, and must then protect his chicks from dingoes and wedge-tailed eagles. The precocial chicks follow him for six months, but find their own food.*
Photo: Gary Unwin, Shutterstock.com

Mute Swan (*Cygnus olor*)

A very large water bird with white plumage, the mute swan is the most common swan. A native of northern Eurasia and North Africa, it has been introduced and is now well established in North America, where it lives in marshes, ponds, lakes, and sheltered marine estuaries and bays. It has a long neck and short black legs and feet, and is entirely white except for its orange bill with a black base and the black knob over the bill. Mute swans measure about 60 inches (1.5 m) long from bill tip to their feet, and weigh up to 30 pounds (13.5 kg), making them one of the world's largest flying birds. Their wingspan measures up to 98 inches (2.5m), and they fly with their long neck extended. They eat mainly water plants and invertebrates, especially water snails.

Mute swans are monogamous but do not pair for life, rather forming a pair bond for one season or perhaps longer. They make a large pile of vegetation on which to lay their four to six eggs, and may use a muskrat mound as a nest. Their incubation period averages 35 days, and the parents share the duties and are extremely aggressive; the bird patrolling their territory attacks anything that gets too close. Even mallard ducklings have been killed when they innocently approached a mute swan nest. Their cygnets have a covering of thick brownish-gray down and are the original "ugly ducklings," attaining their white plumage at the age of one year, when they are driven away by the parents as they prepare to nest again. When they are small they may be carried on a parents' back.

Wattled Crane (*Bugeranus carunculatus*)

This is the largest of the six species of African cranes, a dark blue bird with gray wings, white breast and neck, red face, and long feathered wattles beneath its chin. It stands 53 inches (1.3 m) tall. It is dependent upon wetlands, now increasingly rare due to human disturbance and the draining of marshland for agriculture. It lives mainly in central-southern Africa, from Tanzania to Botswana, but there are two small isolated populations in southeast Africa and in Ethiopia. It is most plentiful in the large river basins of the Zambezi and Okavango. Mainly a vegetarian, it prefers the tubers and rhizomes of aquatic plants, plus snails, but also visits cultivations to feed on grain and insects. The wattled crane is monogamous and very territorial. Its nest is a large pad of grass and sedges, usually in shallow-water marshes but occasionally in drier grasslands, where there is a high risk of loss from grass fires. It normally lays a single egg, and even when two are laid only one chick is raised. The incubation period of 34 days is shared by both parents, and they also provide care until the chick fledges, a lengthy period of up to four months during which time it is very vulnerable to predators—terrestrial carnivores and large raptors. Both parents feed their chick, presenting food items held in the tip of the bill and dropping them in front of it, and it eventually follows them to a food source.

Ring-necked Pheasant (*Phasianus colchicus*)

A stocky game bird with a long tail and rounded wings, small head, and thin neck, the "ring-neck" reaches a length of 30 inches (76 cm). There are many wild subspecies of this pheasant, and considerable hybridization has occurred during its domestication. Its normal coloration is a bright green head, bright red facial skin, and the distinctive white ring around the neck. The body plumage is basically golden with black spots and with iridescent blue and green feathering. The tail feathers are golden, long, and pointed. A native of eastern Asia, the ring-neck was introduced into California in 1857. As a result of natural range extension and many other introductions, it is now a common game bird in many central and northern states and is the state bird of South Dakota. Although a bird of the grasslands and farmland, it requires "cover" in the form of hedgerows or patches of woodland. The male ring-neck establishes a territory and attracts several females, which are mated and then nest within his area, so he is considered polygynous—having more than one partner. He provides "protection" from other male intruders, but the hens incubate their own eggs and raise their chicks without his help. The average clutch of 12 eggs is incubated for 24 days, in a nest of grasses hidden in the undergrowth or tall grass. The precocial chicks are covered with down and their eyes are open when they hatch, and they are initially shown what to eat by their mothers, in typical gallinaceous bird manner. She cares for them until they are six weeks old, brooding them at night and during inclement weather.

Pectoral Sandpiper (*Caladris melanotus*)

The pectoral sandpiper is a streaked brown "shorebird," about 8 inches (20 cm) long, with a buff breast, white underparts, and a black patch on the rump. It has greenish-yellow legs and a thin dark bill. It breeds on the Arctic tundra but winters in South America, making a journey of almost 10,000 miles (16,000 km) twice annually in the spring and fall. The harshness of life on the tundra, with food at a premium during the short Arctic breeding season, has resulted in some most unusual behavior by this species. The males arrive on the breeding grounds before the females and establish territories. When the females arrive, the males attract them with a flight display, rhythmically expanding and contracting the air sacs in their breasts. They are promiscuous; males mate with multiple females and females mate with several males. Each female then builds a nest—usually a well-hidden scrape lined with grass and leaves. The males then leave for South America before the eggs hatch so they do not compete for food with the hens and their chicks. The females provide all the parental care, at least during the incubation period of the four eggs, which lasts 21 to 23 days. The chicks are almost super-precocial as they leave the nest and feed themselves soon after hatching, but their mother stays close by and broods them when necessary. She stays with them for up to three weeks, by which time they are starting to fledge, and she then abandons them and flies south. The chicks are capable flyers at the age of four weeks and then also migrate.

Notes

1. The exception is the African black coucal (*Centropus grilli*), in which a female mates with more than one male and lays eggs for each to incubate and then raise the altricial chicks.

2. There are several exceptions. Megapode chicks and black-headed ducklings are totally independent when they hatch. When just two days old, ancient murrelet chicks leave their nests and accompany their parents to sea, where their waterproof down protects them. Also, flightless species such as the flightless teal and steamer ducks have difficulty covering their brood with their abbreviated wings.

Mandrills At birth, baby mandrills have a coat of hair and their eyes are open, and they cling tightly to their mother's belly when she walks or climbs. Mothers have a long bond with their offspring. If this baby is a female, it will stay with her into adulthood; if a male, it will leave when it reaches sexual maturity, and will eventually become even more colorful than its mother.

Photo: Eric Gevaert, Shutterstock.com

Discus Fish and fry   *Discus fish fry or hatchlings shoal around their mother, for the first few days of their lives eating solely secretions from her skin. This species is the only fish known to feed its offspring.*
Photo: Courtesy Glenn Thode, Gwynnbrook Farm

Eastern Gobbleguts   *A small (4 inches/10 cm) marine mouthbrooder of the sheltered coastal waters of Queensland and New South Wales, the gobbleguts has a wide, oblique mouth into which it crams its eggs after they have been fertilized by its mate. They are held there safely until they hatch.*
Photo: Courtesy Ákos Lumnitzer

American Alligators   Three baby alligators rest on their mother's back. After assisting their hatching, and then carrying them to water, she watches the crèche of babies for their first winter. Their many predators include large fish, other alligators, raccoons, and storks; and when she sounds the alarm by vibrating her body, the hatchlings clamber aboard for safety.

Photo: Michael Thompson, Shutterstock.com

Carpet Python   A carpet python coils around her eggs to incubate them. Of all the ectotherms, or cold-blooded vertebrates, only the pythons are known to generate warmth, in their case purely to aid the hatching of their eggs. They do this by "shivering"—making rhythmic muscle contractions—every few seconds to raise their body temperature several degrees above the ambient temperature.

Photo: Courtesy Will Bird

Least and Royal Terns *A breeding colony of least and royal terns on a beach in North Carolina. These birds do not make a nest, but simply lay their eggs on the sand. Note the eggs spaced exactly apart in the foreground.*
Photo: Courtesy Shannon Kondrad. USGS, Patuxent Wildlife Research Center

Atlantic Puffin *A puffin arrives back at its nesting burrow with several small fish caught after an underwater chase. Its round tongue and serrations in its upper mandible allow it to hold many fish crosswise in its beak. Its single chick is covered with soft down when it hatches, but the parents must brood it for the first week to keep it warm.*
Photo: Joe Gough, Shutterstock.com

Red-necked Grebes   Grebe chicks are precocial and usually leave the nest soon after they hatch. Although they can swim and dive, they often hitch a ride on their parent's back. They are offered food by the parent in its bill.

Photo: Courtesy Donna Dewhurst, USFWS

Blue-footed Booby   Unlike practically all birds, the boobies or gannets lack a brood patch —the highly vascularized skin that warms the eggs during their incubation. They hold their eggs under the webs of their large feet to incubate them, and as they near hatching they rest them on top.

Photo: Rebecca Picard, Shutterstock.com

Brush-tailed Possums   An Australian marsupial, the brush-tailed possum's baby is born after a gestation period of just 17 days, and spends five months in its mother's pouch. Then it climbs onto her back and rides for two months, by which time it really does seem too large to be carried.

Photo: Polina Yun, Shutterstock.com

Japanese Macaques   With the northernmost range of all primates, Japanese macaques must cope with harsh winter conditions in their mountainous homeland. Within a few months of its midsummer birth, the baby "snow monkey" faces subzero temperatures, snow, and cold winds, and is held tight by its mother, already in her thick winter coat.

Photo: WizData.Inc, Shutterstock.com

Klipspringer   A male klipspringer stands guard while his mate nurses their baby, a "hider" that spends up to three months hidden in the undergrowth. One of the few species of monogamous antelopes, in which a male and female form a lasting pair bond, he can be certain the baby is his, so it is in his interest to help it survive.

Photo: Courtesy Ralf Schmode

Nyalas   A pair of nyalas with their calf drink at a muddy water hole. An antelope of southeastern Africa, the nyala shows considerable sexual dimorphism—obvious differences between the male and female. Initially the calf is an image of its mother, but its sex will determine its eventual size and color.

Photo: Photos.com

Dromedaries   A dromedary, or single-humped camel, nurses its newborn calf. Born after a gestation period of 13 months, it can follow its mother the day after its birth and will be nursed for about one year. The dromedary's milk composition is very similar to regular cow's milk, but the fat content is considerably reduced when the camel drinks her fill after several waterless days of desert travel.

Photo: Paul Prescott, Shutterstock.com

Domestic Pig and Piglets   Domestic piglets are sparsely haired at birth and must be kept warm until they can thermoregulate, which happens when they are four or five days old. They are mobile soon afterwards and can follow their mother, stopping frequently for a drink wherever they are.

Photo: Photos.com

6 An Introduction to Mammals

Reproduction in the higher vertebrates or amniotes is characterized by the presence of the amnion, a thin but tough membrane that encloses the embryo, and later the fetus, enveloped and protected by the amniotic fluid. In many reptiles and all birds it is contained in a shell, and a few mammals also still lay shelled eggs. They are the monotremes, among the first mammals to arise from the reptiles during the Mesozoic Era about 185 million years ago. But only three species of this early link between the reptiles and mammals have survived to the present day—two echidnas and the duck-billed platypus. The other mammals that began to evolve at the same time long ago were more successful, and soon diverged into two groups—the marsupials, or pouched mammals (*Metatheria*); and the placental mammals (*Eutheria*).

The marsupials evolved beyond the reptilian egg-laying stage but did not develop a placenta to nourish their young in the uterus. Their offspring are therefore born at a very premature stage and are then raised in their mother's pouch. The eutherians were the most successful of them all, for they developed an amniotic membrane that allows the passage of nutrients and wastes between the embryo and the mother. This permits a longer gestation period and the full development of the fetus in the womb, so the young are born in a more advanced state than the implacental marsupials, followed by a shorter period of dependency on the mother. Despite the longer gestation, however, there is considerable variation in the condition of the higher mammals at birth, ranging from the highly precocial ungulates such as the deer and antelope, which can stand and run soon after birth, to the blind and helpless tiger and lion cubs, which are said to be altricial. These variations occur even within related species; hares have well-furred and mobile babies, born with their eyes open, while in contrast young rabbits are naked, blind, and helpless.

The extinction of many reptiles at the end of the Mesozoic Era about 165 million years ago allowed these mammals to flourish, and they evolved into the current seven orders of marsupials and 16 orders of placental mammals. The exact relationship between the three major groups of mammals, however—especially between the egg-laying monotremes and the others—is still debated. But the fossil evidence does not support the original belief that the marsupials were primitive forerunners of the placental mammals, and it is now thought that both developed separately toward the end of the era. There is no doubt, however, that the placental mammals were more successful except in Australia, where they were absent until quite recently, and which allowed the marsupials to flourish there.

Parental care in the mammals is dominated by the regular supply of milk. Whether they lay eggs like the echidnas, are implacental kangaroos or placental rats and mice, all mammals suckle their young, but their caring behavior varies considerably. In some placental mammals, the father and siblings, and even other members of the social group, may also be involved in caring for the young; but only the females can suckle them, despite the occasional claims of male animals producing milk. Lactation by male mammals is physiologically possible, however, and has been reported in domesticated animals and in humans. Biologists collecting bats in Krau Game Reserve in Pahang, Malaysia, recently caught some male Dayak fruit bats (*Dyacopterus spadiceus*) that had enlarged nipples from which a small amount of milk could be expressed, but there are no accounts of males having sufficient milk to raise babies.

Nursing allows the baby mammal to continue its development outside the womb, and in the altricial species frees the mother from the constraints of carrying a large litter to its full independent development, and thus increasing her vulnerability to predators while doing so. But there are costs and risks associated with this form of reproduction. Lactation is a costly aspect of parental care in mammals, and returning to the same place regularly to warm or feed the litter also increases the mother's vulnerability to predation.

Mammary glands and milk production, like hair, are uniquely mammalian. Mammals were the first animals to produce real milk, as opposed to the secretions of certain birds that have been called milk. The females of all the current 4,700 species produce milk, without which their newborn young would not survive. The glands are highly developed sweat glands, comprising a system of ducts surrounded by glandular tissue. It is uncertain when they first developed, but in the surviving monotremes they are simple collections of glandular tissue in the wall of the abdomen. Because they are not associated with teats or nipples, the milk is discharged into depressions from where it is licked by the baby echidna or platypus. The marsupials and placental mammals developed a more direct way to supply milk to their offspring, via teats or nipples. The marsupials actually have a very sophisticated milk supply and delivery system, especially in the compositional changes during lactation, and the ability of different teats in the same animal to supply different milk. In the pouched marsupials, the teats are always inside the pouches, whereas in the higher mammals, the mammary glands, and thus the teats or nipples, are paired and run in "milk lines" along the belly. Many, such as pigs and dogs, have numerous pairs, others a single pair, that may be situated in the armpits (in the manatee), pectorally (in the apes), or abdominally (in horses). Mammary gland

Bonnet Macaques *A mother bonnet macaque holds her baby. A common Indian monkey that lives in a multi-male, multi-female, or promiscuous social system, the mother is totally responsible for the care of her altricial offspring. It will begin eating solids at the age of eight months and will likely be weaned when about 15 months old. Babies carried in this manner can suckle almost on demand, and their mother's milk is therefore quite weak.*
Photo: Vishal Shah, Shutterstock.com

growth is influenced by the hormones progesterone, prolactin, and oxytocin, each of which also plays a role in birth and the production and flow of milk.

Milk is the nutrient fluid produced by the mammary glands, a complete food and the sole food for baby mammals for several weeks in the smaller species and many months in the larger ones. It contains varying proportions of all the components essential to life and growth—protein, fats, sugars, vitamins, minerals, and water. With few exceptions[1] all mammalian milk contains exactly the same basic ingredients, although the actual composition varies according to the species and its lifestyle. A mother's first milk is called colostrum. It contains maternal immunoglobulins (antibodies) and enzymes that counter infections of the digestive tract and assist in developing the infant's immune system. In the placental mammals, it is produced for only two or three days after birth as the baby's absorption period is very short, because the development of enzymes degrades the antibodies. Some mammals—for example the primates and rabbits—acquire their maternal antibodies while in the uterus.

Water is the major component of most milks, as it is essential for the neonate and provides the suspension medium for the other components. Cow's milk contains about 86 percent water, and horse milk is even more dilute, having only 11 percent solids (fat, protein, and lactose) and therefore 89 percent water. The pinnipeds (seals and sea lions) are exceptions, as the solids in their milk exceed its water content. Milk fat, or butter fat, is the most variable constituent of milk (both between species and throughout lactation periods) and can be affected by an animal's diet and its condition. It is the infant's energy source, and is used to form adipose tissue as most mammals are born with very little.

Several types of proteins or caseins occur in mammal milk, and they are involved in growth, disease resistance, and nutrient transportation. In the baby's stomach the protein molecules are converted to curds by the enzyme rennin. Lactose is the main carbohydrate in placental milk, and is digested by the enzyme lactase to provide an easily assimilated source of energy for the suckling baby. Milk contains all the vitamins and minerals required by the baby. Fat-soluble vitamins A, D, E, and K occur in milk fat, and water-soluble B vitamins in the water. Calcium and phosphorus are the major minerals in milk.

The actual composition of milk—the percentages of the various components— varies considerably. The protein content is low in primate milk, (averaging just 2 percent in Old World species, but over 4 percent in Neotropical ones), but it is high in carnivores, with domestic cat's milk containing 11 percent protein and dog's milk having 9.5 percent protein. There is very little fat in the milk of the horses and asses, but cat's milk has 10.9 percent fat, dog's milk has 8.3 percent fat, bear's milk has 30 percent fat, and seal milk contains up to 60 percent fat. Whole milk from the supermarket has only 3.5 percent fat and slightly less protein, whereas the domesticated water buffalo's milk (from which gourmet mozzarella cheese is made) contains 10 percent fat and 5.8 percent protein. Lactose or milk sugar varies even more, with just under 5 percent in domesticated cows, a little more in horses and asses, but is absent in the California sea lion and the monotremes and occurs as only a trace in the marsupials. Changes in composition generally occur during a lactation period. During the domestic dog's 45-day lactation, for example, the protein and fat increase in the middle of the period, then drop back, and there is a large increase in the calcium content. Marsupial milk changes dramatically in composition during the babies' pouch life.

The period of dependency on milk varies from just a few days for some seals to many months for young elephants, but a few weeks of nursing is normal for most altricial babies. The almost continual breeding cycle of many species means that the young of the previous litter are denied the opportunity to continue suckling, but animals such as lions and bears, in which survival depends on an extended period of care, allow their cubs to nurse for much longer.

All mammals provide parental care for their young; and initially it is the mother's responsibility, for only she can supply the nourishment essential for her offspring at birth, and for most species for some time afterward. These constraints, immediately following a lengthy gestation period, affected the evolution of parental care in mammals. They are mostly polygamous, and while the females are caring for their offspring, the males are free to find other mates. But in some species, even during the suckling period, the fathers also provide care, increasingly as the infants grow. In the monogamous species, the father may also care for his offspring. When his help is not absolutely necessary in raising the young, it is called facultative care; and this occurs in the duikers, prairie voles, elephant shrews, and coyotes. When the father's help greatly improves the chances of reproductive success, or for some animals may be absolutely necessary, it is called obligate care. The douroucouli is an example of this behavior, in which the father carries the baby and hands it over only for nursing. To a lesser extent the klipspringer father's actions during the raising of the calf also constitute obligate paternal care, as he stands guard while his mate is nursing their baby, hidden in the undergrowth. As inconsequential as paternal care may seem, it is known to improve the infants' chances of survival. Obviously, it has some value or it would not have evolved in the species that practice it.

Polygamous behavior in mammals takes the form of polygyny, in which the male seeks more than one mate, and the much rarer mating arrangement called polyandry, in which a female mates with more than one male. The most well-known examples of the latter are the cooperative breeders, the marmosets and tamarins, which generally live in small groups comprising one female and two mature males with their offspring; and the naked mole rat, in which a "queen" mates with several consorts. In the social, polygynous mammals, the zebra stallion controls a herd of mares and their foals, and the bull elk has a "harem" of cows. Sexual dimorphism is generally synonymous with polygyny, the males being much larger than their mates, which is quite obvious in the huge elephant seal courting his considerably smaller females. There are less obvious examples of polygynous behavior, however, such as the black rat and house mouse and the black bear and polar bear, in which dominant males mate the females within their territory; and the African lion, in which the alpha male mates the females in his pride.

The type of parental care in mammals, as in birds, depends upon the degree of development of the young at birth, specifically whether they are precocial or altricial. Newborn altricial young are basically helpless, only able to wriggle about in their nest or den. They may be naked like the hatchlings of the monotremes, the pouch young of the marsupials, and the young of placental species such as rats and mice; or they may be furred like cougar cubs or wolf pups. Their eyes and ears are closed when they are born, they cannot regulate their body temperature, and cannot survive without the external warmth and protection provided by the nest or den, their siblings, and most importantly their parents.

Altricial young are either raised in a pouch, like the egg-laying monotremes and most of the marsupials; or in a den or nest, like the young of many placental mammals such as the carnivores, most rodents, rabbits, some insectivores, a few primates and the swine—the only ungulates or hoofed mammals to make nests. Little is known about the commencement of thermoregulation in wild altricial

Capybaras   *The world's largest rodent, the capybara's babies are precocial at birth. Covered with hair and able to thermoregulate, they are mobile and can swim, so they follow their mother, and suckle almost on demand. They eat solids within a few days of birth, but continue to suckle until they are four months old.*
Photo: lee torrens, Shutterstock.com

species, except for the pouch babies of the tammar wallaby, which cannot thermoregulate until they are six months old. Domesticated kittens and puppies begin to regulate their body temperature when they are three weeks old. Young animals produce heat through the breakdown of brown fat, present on the upper back and shoulders of newborn placental mammals and packed with mitochondria that are the principal energy source of the cells. In altricial prey species such as rats, mice, and hamsters, the young usually depart soon after being weaned, as their mothers must prepare for another litter. In social species, such as crested porcupines and pigs, the young stay with the group. In predatory altricial species the young generally have an extended period of maternal care, as they must also be taught to hunt. Tiger cubs stay with their mothers for at least two years and polar bear cubs even longer, up to 30 months. Tiny least weasel babies, weaned at the age of five weeks, hunt with their mother until they are 10 weeks old.

In total contrast, precocial young are born with their ears and eyes open, they have a coat of fur or hair, and they can thermoregulate. Most important, however, they are mobile, to the extent that they can keep pace with their mother and the herd soon after birth, or can be left on a beach or an ice flow. With few exceptions, precocial species are prey animals that provide flesh for the larger carnivores, both marine and land dwellers. The hyenas are the only carnivores whose young are well developed at birth, and may be considered precocial, yet they are raised in dens just like altricial babies.

Between these two definite types, there are the semi-precocial species; animals that carry their babies. Most are furred at birth, their eyes and ears are open, and they are probably able to thermoregulate, to a degree. They can cling to their mother, and may also be held by her, and can reach a nipple and feed without assistance, but away from their mothers they are helpless. They have been called precocial, but they hardly warrant that designation.

Consequently, there are five major forms of parental care in the mammals—three altricial and two precocial. The altricial species are the egg-laying monotremes and the pouched marsupials; the helpless young mammals left in a nest or den, and the babies that are carried everywhere by their parents. Precocial species are those that follow their mother soon after birth; and the baby pinnipeds, furred and maintaining their own body warmth, that wait on a beach or pack ice for their mother's return from the sea. In each group the type of care varies considerably, in form and duration, and their milk differs in its energy content. The following chapters chronicle these varying types of care. The format has no taxonomic relevance, and is organized entirely from the point of view of parental care.

Notes

1. The milk of the monotremes and the California sea lion lacks lactose, and that of most marsupials contains only traces, but it is the main carbohydrate in the milk of the higher or placental mammals.

7   Pouch Babies

The first mammals to evolve from the reptiles in the Mesozoic Era about 185 million years ago were the *Prototherians* and the *Therians*. The *Prototherians* were not successful and died out except for three species—the duck-billed platypus and two echidnas—that still lay eggs like their reptilian ancestors. Consequently they are the most primitive mammals alive today, relicts with reptilian and mammalian features. The *Therians* were more successful and eventually diverged into two groups, the marsupials (*Metatheria*) and the placental or higher mammals (*Eutheria*). Marsupial evolution did not include the development of a placenta like the higher mammals, and their offspring are therefore born prematurely after a brief gestation period. Most of the present-day survivors of these first mammals have pouches or marsupiums. The platypus is the only monotreme without a pouch; some marsupials also lack pouches, while others develop rudimentary ones, little more than flaps of skin, just for the breeding season, and their babies simply dangle from the teats. Generally, however, the premature babies of these animals are raised in pouches protected from the environment almost as securely as if they were completing their development in their mother's uterus. Both groups produce milk and suckle their young.

■ MONOTREMES

The monotremes are the most primitive mammals, combining reptilian and mammalian characteristics. They are restricted to Australasia and are the only living links between the reptiles and mammals. Their survival resulted from the splitting of Gondwanaland some 80 million years ago, allowing the animals on the southern land mass to continue their evolution in the absence of the developing higher or true mammals in the rest of the world. The monotreme's ancestry is unclear mainly because they are toothless, and the classification of early mammals is based largely on fossilized teeth. The general belief, however, is that they diverged from the marsupials and the early placental mammals long ago. Their oldest known fossils, from Argentina, are 110 million years old, so they began their development there before the breakup of Gondwanaland. Their skeletons and muscles bear similarities to the moles, and it has been suggested that they evolved from fossorial or burrowing animals.

Although they are indeed mammals and are therefore warm-blooded, the monotremes have not completely mastered the ability to thermoregulate, and their temperature fluctuates with the environment. They have a low metabolic rate, and their body temperature is much lower than most mammals. Like the reptiles and birds, they have a cloaca, the single opening for passing fluid and solid wastes, for mating and for laying eggs. They also resemble the reptiles in the structure of their eyes and in certain aspects of their skulls, ribs, and vertebrae. Like the higher mammals, however, they have a coat of hair, or hair modified as spines, they have mammary glands for milk production, and they suckle their young. They were therefore the first animals to raise their young with milk.

Monotremes have a typical mammalian four-chambered heart, whereas most reptile's hearts have three chambers. Their babies have teeth but lose them as they mature, and the adults are toothless and crush their food with their hard palates and tongues. At least one of the species—the short-beaked echidna—hibernates, and the duck-billed platypus may become torpid in cold water. Male echidnas and platypuses have horny spurs on their ankles that are connected by ducts to poison glands in the thigh. The glands and ducts of the echidnas, however, are vestigial and do not function, unlike those of the platypus whose spurs are dangerous to humans.

Monotremes are the only mammals to lay eggs. They resemble those of the reptiles, with leathery skins and large yolks. There is some passage of nutrients from the uterus into the egg's embryo via the extra-embryonic membrane called the allantois, and through the yolk sac placenta prior to laying. The embryo therefore begins its development before the egg is laid, similar to the early stages of viviparous reptile reproduction, in which the mother contributes to the development of the egg through the passage of uterine nutrients. As they lack a real placenta, however, they do not pass birth fluids or membranes when they lay their eggs. The eggs are incubated outside their bodies in the manner of the birds and the oviparous reptiles, but the actual external incubation period is quite short, usually about 10 days; and baby monotremes are tiny, naked, blind, and helpless when they hatch. They are the only mammals that can truly be said to hatch rather than to be born.

They have resembled reptiles to this point, but with the hatching of their young the monotremes almost "become" mammals, for they suckle their young in a way. They have well-developed mammary glands, but no teats like all other mammals, even the marsupials; and their babies lap the milk as it is expressed from areolae. Despite being extremely well adapted and successful at what they do, only three species of monotremes have survived—the duck-billed platypus and two species of echidnas or spiny anteaters. Males are believed to mate several females, so they are considered polygynous, but they are solitary animals that go their own way after mating, and there is no opportunity for fathers or other family members to help raise the babies. Like the marsupials, the female monotremes are therefore totally responsible for raising their young.

Echidnas or Spiny Anteaters

Echidnas resemble large hedgehogs, and for many years just two species were recognized, each in its monotypic genus. The short-beaked echidna (*Tachyglossus aculeatus*) has a compact muscular body covered with coarse hair and barbless spines, supported on short bandy legs and big feet with large claws. It occurs in a wide range of habitat, in forest, grasslands, rocky hillsides and sandy areas in Australia, Tasmania and southern New Guinea. It reaches a length of 18 inches (45 cm), with a tail 3 inches (8 cm) long, and weighs up to 13 pounds (6 kg). Its spines are just over 2 inches (5 cm) long and are yellow with black tips, except for the Tasmanian race (*T. a. setosus*), in which they are much shorter and, in some individuals, almost concealed by their very thick coat, which provides protection in the cold winters. Both the male and female have spurs inside their ankles, but unlike the male platypus, the glands and ducts are no longer functional. Despite their spiny covering, dingoes are able to kill them.

Although in a recent revision of the other echidna genus, *Zaglossus*, it was suggested that three species should be recognized, most taxonomists still believe that there is just a single one—the long-beaked echidna (*Zaglossus bruijni*) of the moist mountain forests and meadows up to 13,000 feet (4,000 m) in New Guinea. It is a larger animal than the short-beaked echidna, reaching a length of up to 30 inches (75cm) and weighing 30 pounds (13.6 kg) making its tiny head appear even smaller by comparison. With its very long downward-curving snout, it resembles a huge weevil. Its spines are short, usually grayish white, and are sometimes almost hidden on the back by thick black hair. Like its smaller relative, it also eats ants, but earthworms and beetle larvae form the bulk of its diet. To aid the capture of worms, it has tiny backward-projecting spines on the end of its tongue with which it hooks them into its mouth.

Echidnas have short and sturdy limbs, broad feet with strong claws, and are powerful diggers. Their feet turn sideways when they walk to protect the long claws. Their small eyes are situated at the base of the snout and look ahead, providing binocular vision although not very acute sight. Food is located by smell, aided by chemical-sensing vomeronasal organs, and possibly also with the help of the electroreceptors in the snout, which are believed to detect the electrical signals given off by its prey. The snout is a remarkable organ, with receptors also for temperature and touch, and with great sensitivity to ground vibrations. It is quite tough and is used to dig into rotten wood and soil, and the small mouth at the end of the snout opens just enough for the long sticky tongue to flick out to gather ants. Enlarged salivary glands keep the tongue moistened. Lacking teeth, the echidna relies mainly upon horny ridges on the palate and tongue to grind its food. Echidnas are insectivorous and cannot cope with large food items. Their diet comprises mainly of ants and termites, plus earthworms and beetles; but they cannot digest chitin, so the exoskeletons of beetles are passed out in their feces.

Echidnas have good hearing, even though their external ears are almost hidden by hair. Like their reptilian ancestors, they have a chamber called the cloaca into which the digestive, excretory, and reproductive systems all enter, opening to the outside via the anus. Despite their appearance, echidnas can run fast, are good climbers, and can swim; but if they cannot escape by running or burrowing into soft soil, they roll into a tight ball. They are nocturnal and shelter in burrows, hollow logs, or rock crevices, in which they wedge themselves tightly with feet and spines; but they are often out during the day, even when it is sunny. They have a good sense of smell, but their sight and hearing are not well developed.

The echidnas are poor thermoregulators and must avoid extremes of temperature. They burrow into ant mounds or hide in caves or crevices to escape cold and heat, and become torpid during very cold weather. Their body temperature fluctuates widely, between 82.4°F (28°C) and 91.4°F (33°C) compared to the marsupial's 95°F (35°C) and the higher mammal's average of 98.6°F (37°C). Therefore, they are considered heterothermic, unable to maintain a set temperature, and are the most "cold-blooded" mammals after the tailless tenrec and the two-toed sloth. In the colder parts of their range—southeastern Australia and Tasmania—echidnas cannot keep warm in winter and become inactive and torpid, their body temperature dropping to 39.2°F (4°C) for extended periods.

Echidnas are solitary animals that socialize only at mating time. They nest in an old wombat or rabbit burrow, in a crevice, beneath a log or under a rock overhang. The female lays a single egg—the size of a small grape with a rubbery shell—about 23 days after mating. During its development within her body, it is provided with nutrients via the uterus wall. The echidna's stomach pouch is a temporary one, developed just for the breeding season, and is little more than a depression with muscular walls. It is unclear how the egg gets into the pouch, as the mother certainly cannot pick it up. She may curl around and lay her egg into or close to the pouch entrance, or she may maneuver the egg in with her snout. The egg's incubation period in the pouch is 10 days, and the hatchling is little more than an embryo, about 5/8 inch (15 mm) long, naked, blind, and helpless, but it must wriggle about in the pouch to reach the milk patches.

Despite their unusual mix of reptilian and mammalian characteristics, the echidnas have a well-developed milk production system, but their method of delivery has not reached the higher mammal's degree of perfection. Echidnas lack teats and the infant laps the milk secreted by glands opening onto two areas on the mother's belly within the pouch, known as mammary patches. Echidna milk is very dilute (low in solids) initially, but soon changes to a very rich milk containing 30 percent fat, 12 percent protein, and a small amount of sugars. The milk lacks lactose, unlike placental mammal milk, in which the main sugar is lactose; and it differs from platypus milk mainly in the structure of the other carbohydrates. It is rich in iron, presumably to compensate for the baby's tiny liver.

The young echidna, known as a puggle, stays in the pouch until its spines begin to grow when about 50 days old, and is then left in the nest when the mother goes outside to eat. She apparently returns infrequently, perhaps only every two days, to nurse it. The puggle continues to suckle for several months, its mother assisting by pushing it under her body and then arching her back so that the pouch is clear of the ground, and the infant can cling upside down to her fur and put its head into the pouch. At the age of six months, the young echidna leaves the nest for short periods and begins to eat termites, and is then very vulnerable to predators, especially feral cats. It is independent when nine months old.

Duck-billed Platypus

The duck-billed platypus (*Ornithorhynchus anatinus*) is the world's most unusual mammal. It has the streamlined body of an otter, short and stout limbs with broad webbed feet, a tail like a beaver, and a leathery snout resembling a duck's bill. It has a blackish-brown coat of short and dense fur, composed of woolly underfur and blade-like guard hairs; and pale-yellow or brownish-white underparts. It is active from dusk to dawn, in the rivers and ponds of eastern Australia from Victoria to central Queensland, and in Tasmania. Despite its strange appearance, it is a very successful animal, for fossils show that it looks the same today as it did many millions of years ago. It is so unique, it is the only species in its genus.

The platypus is an excellent swimmer and diver, and mainly uses its forefeet for propulsion. As it is a mammal, it cannot breathe underwater and must return to the surface regularly for air, just like an otter. It is a very buoyant animal, and to stay submerged when it stops swimming, it wedges itself beneath a submerged log or rock to prevent bobbing back to the surface. The platypus is entirely carnivorous and probes in the river and lake bottoms for aquatic invertebrates, freshwater crustaceans and snails; it also catches small fish, tadpoles, and frogs, holding its prey in cheek pouches until it returns to the surface. As it lacks teeth, horny plates in each jaw, ridged in front and smooth at the rear, crush the platypus's food. These plates are continually growing as they are worn down by grit scooped up with the prey.

Male platypuses have hollow spurs on their ankles, connected to poison glands, and are the only mammals able to actually inject venom, unlike the mildly venomous shrews that must chew on their victim to force venom into the wound. Although all baby platypuses have spurs, the females lose them as they mature. The venom can kill a dog, and while obviously a danger to humans, it is not known to be fatal. Their major predators are carpet pythons, large monitor lizards and water rats that kill the babies in the nest, and foxes and feral cats that kill juveniles when they leave the nest burrow.

The platypus' duck bill is a very sensitive organ, and is swept from side to side to locate its food. In addition to its tactile sense, the bill also functions as an electro-receptor, detecting the weak electric fields emanating from small animals and locating them in the mud and under rocks. The special olfactory receptors known as vomeronasal organs, which are present in many reptiles, are also well developed in the platypus. They contain sensory neurons similar to those in the nose that detect chemical compounds, including pheromones—which carry messages between animals of the same species—and are connected to the olfactory nerves. Platypuses lack outer ears or pinnae, so hearing is not their best sense. Under water, their ears and eyes are covered with folds of skin anyway, so they are effectively blind and deaf and rely completely on their sense of touch.

Platypuses dig burrows in river banks, the entrances to which are just wide enough to enter and squeeze the water from their coats before they reach the nest. Their burrow entrances are always several feet above the water level, but may be submerged by floods. A mated pair generally has two active burrows, one that is normally used by the pair, and another just by the female as a maternity burrow, for only the female platypus cares for her young. The nest burrow may be 50 feet (15 m) long, with a nest of damp grass and leaves at the end. Platypuses breed in the spring, mating in water, and one to three eggs are laid about two weeks later, after being supplied with nutrients from the uterus wall. The size of house sparrow's eggs, the female curls around them, holding them with her tail against her belly to provide the correct temperature. The actual incubation period is only about 10 days, and the hatchlings are tiny, blind, naked, and helpless. They are brooded and suckled for four months before they appear fully furred at the burrow's entrance.

Platypus milk oozes from glands on the mother's abdomen that open at the base of mammary hairs—the forerunners of nipples or teats—and is licked up by the babies. She has no pouch, and lies on her back with the babies resting on her stomach to feed. There is no evidence that platypus milk changes in composition during the lactation period, but this may be due to the relative lack of available samples. During the middle of the lactation period, platypus milk contains 22 percent fat, 8 percent protein, and 3.5 percent sugars. But lactose is absent, as it did not evolve until later, with just a trace in some marsupials, followed by its development as the major sugar in the milk of higher mammals. The young are lightly furred at six weeks and their eyes are open, but they do not leave the burrow until they are weaned when about five months old. Young platypuses have teeth, but lose them by the time they leave the nest tunnel, when they are replaced by the keratinous grinding pads.

■ MARSUPIALS

The marsupials (*Metatheria*) follow the monotremes as the second most primitive order of mammals. They are the pouched mammals, in which the female typically has a pouch or marsupium, although it is very rudimentary or even absent in some species. Their common linking factor is the lack of a placenta, through which the higher mammals nourish their embryos in the womb; marsupial young are therefore born in a very premature state, and their development continues in the mother's pouch. Consequently, they are more correctly known as implacentals, and are characterized by short gestation periods and long suckling periods; whereas the higher mammals have long gestation periods, and most species give birth to large and well-developed young that are soon weaned. The marsupials occupy the position zoologically between the egg-laying monotremes (*Prototheria*) and the higher placental mammals (*Eutheria*).

The marsupials evolved in North America, traveled down to South America and then into Australia, when all the land masses were attached as Gondwanaland. They were originally believed to be the ancestors of the placental mammals, but the fossil evidence does not support this. Although some continued to evolve in the New

World, they only really flourished in Australia, due to the absence of the evolving carnivores. Those left behind in South America, after its separation from Africa and Australia, did not have such an easy time, for they had to contend with the evolving placental mammals, including cats, dogs, skunks, and raccoons; but several species survived, although they are little changed from their ancestors. It is significant that they are small, mostly nocturnal, and are experts in the art of daytime concealment. They are mainly arboreal and many have prehensile tails, but the most aberrant species (that deviates in important characters from its nearest allies) is the yapock (*Chironectes minimus*), a semiaquatic animal that can close its pouch opening tightly when it enters the water. This is especially unusual as most of the New World marsupials lack pouches.

Marsupials have smaller brains than the placental mammals, but although they have always been considered rather stupid animals, laboratory experiments have shown they are as intelligent as many higher mammals. All have marsupium bones, even the pouchless species, that project from the pelvic bone and strengthen the wall of the abdomen to support the weight of the pouch babies. They have a well-developed sense of smell, and mark their territories with scent from their skin glands and with urine and feces. In addition to the standard sense involving the olfactory organs, they also have a vomeronasal system similar to the snakes, in which receptors in the mouth detect chemical compounds and transfer the information via a duct to an organ in the nose, from where it is transmitted to the brain for analysis. Hearing is well developed in many species, especially those with large ears such as the bilby (*Macrotis lagotis*) and the long-nosed bandicoot (*Perameles nasuta*).

Marsupial classification is based on teeth and toes. Species with more than two incisor teeth in the lower jaw (most have four), and many teeth generally, are called polyprotodonts and are carnivorous. Those with just one pair in the lower jaw are called diprotodonts and they are vegetarians, primarily grazing and browsing herbivores. The diprotodonts occur only in Australasia, whereas polyprotodonts live in Australasia and the New World. A further characteristic of the Australian diprotodont marsupials is the unusual structure of their feet, which are said to be syndactylous. The second and third toes are fused together but with the claws separate. The fifth toe (the hallux, or big toe) is missing, and the fourth toe is usually greatly enlarged and clawed—an adaptation for saltatorial or bounding locomotion. The bandicoots are the only polyprotodonts with syndactylous feet.

The American marsupials, 85 species in the family *Didelphidae*, are all small animals. The members of the genera *Didelphis*, *Chironectes*, and *Philander* have pouches, but in all other species the pouch is just an open area with folds of skin on either side of the teats. It cannot contain the young, so they must find a teat which then swells inside their mouth to secure them. All New World marsupials have very short gestation periods, seldom more than 13 days, but even at that age the tiny embryos have well-developed arms and claws and clamber to the teats without assistance from their mother—a rather precarious journey, especially if the mother is disturbed during the process.

Australia is the marsupial's kingdom, where they are the dominant mammals; 195 species, which have evolved into a far wider variety of animals than their

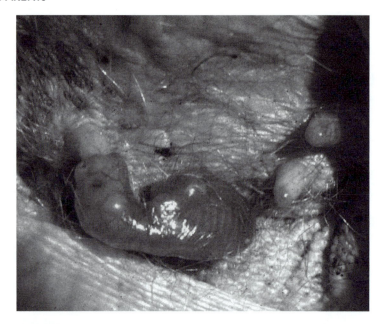

Kangaroo Pouch Baby *After a gestation period of only 35 days, the baby kangaroo crawls unassisted into its mother's pouch and finds a teat that swells and anchors it firmly. Little more than an embryo, it cannot thermoregulate, and will spend many months developing in the pouch, in what is virtually the equivalent of a precocial placental mammal's gestation period.*
Photo: Courtesy Geoff Shaw, http://kangaroo.genome.org.au

ancestors, with both carnivorous (polyprotodont) and vegetarian (diprotodont) forms. The presence of a few indigenous typical Old World species on the island continent, such as bats and rats, results from colonization in more recent times by flying or seaborne invaders from the neighboring Indonesian Archipelago and New Guinea. All other wild mammals in Australia are the result of more recent deliberate or accidental introductions by human agency. Lacking competition from evolving carnivores, Australia's marsupials evolved into an amazing variety of species. They include terrestrial grazers and arboreal browsers, hunters, worm-eaters, honey-eaters and scavengers. The large grazing marsupials have developed a compartmental stomach similar to the higher ruminants (cattle, sheep, and goats) for digesting their fibrous diets by microbial fermentation. Unlike their insectivorous or carnivorous ancestors, the Australian marsupials diverged to take advantage of plant matter in its various forms—grass, leaves, fruit, nuts, flowers, and honey—and developed mouths, teeth, tongues, and digestive systems to cope with their vegetarian lifestyle. These diprotodont marsupials include the wombats, koalas, kangaroos and wallabies, and many species of possums. The polyprotodont or carnivorous marsupials include the marsupial mice, native cats, and the bandicoots. The most familiar marsupials are the kangaroos and wallabies, belonging to the family *Macropodidae,* and therefore usually called macropods.

Parental Care

In the marsupials there is only maternal care, at least while the young are in the pouch, as their development there totally precludes any involvement by the father or others. In the ring-tailed possum (*Pseudochirus peregrinus*), a more social species that lives in small groups, the male may carry the young on his back when they have permanently left the pouch. As females have total responsibility for their pouch young, monogamy is very rare in marsupials. The greater sugar glider (*Petauroides volans*) is monogamous, and a mated pair share their tree hole with several of their young, and the father may also assist when they are out of the pouch. The most common mating system in the marsupials is polygyny, with perhaps a high degree of polygynandry or promiscuity also. Males of the many solitary species mate several females, and it is possible that the females also have several mates. Many of the kangaroos and wallabies are social animals, and polygyny is believed to be their main mating system.

Reproduction in the marsupials is based primarily upon lactation after a very short gestation period, whereas in the higher mammals the development of the young within the mother's body is the major phase, the primary time of investment, that in most species is followed by a relatively short lactation period. Despite lacking a placenta, the embryonic marsupial does receive some nourishment during its brief gestation period in the uterus, via the yolk sac placenta that absorbs nutrients through the amnion. The yolk sac is the membrane-bound compartment of the egg that contains stored food for the developing embryo. In the bandicoots (*Peramelidae*) and the bilbies (*Thylacomyidae*), however, a more sophisticated placenta develops that involves other embryonic membranes, the allantois and chorion, that also provide nourishment to the egg. This resembles the placenta of the higher mammals, but unlike it lacks the villi or tiny finger-like projections.

The marsupial gestation period ranges from a mere 13 days for some opossums to 35 days for the larger kangaroos, with such species as the tiger cats, wombats, and bettongs lying between these extremes with about 22 days gestation each. For the animal's size, these gestation periods are the shortest of all the mammals. After such a brief period of development, despite receiving some nourishment, marsupial babies are literally tiny external embryos. Baby common opossums are only ⅜ inch (10 mm) long and weigh 0.0045 ounces (0.13 g). The young of the largest kangaroos are just ¾ inch (2 cm) long and weigh 0.026 ounces (0.75g).

There are advantages to the marsupial reproduction system. The short gestation period means that the initial cost of reproduction is low compared to the lengthy gestation of the larger placental mammals; but this is countered by the long raising period, so the costs of reproduction are therefore similar. However, if she loses her baby, the marsupial can breed again more quickly than a placental mammal, as she has one "in-waiting" in the uterus. In contrast, the placental mammals must go through the processes of courtship, awaiting estrus, mating, and a long gestation period before giving birth.

Young macropod marsupials are typically called "joeys," and after one is born and wriggles up into its mother's pouch, she mates and conceives again very quickly. About four days later the fertilized egg has reached the blastocyst stage, when it is simply a hollow ball of cells filled with fluid, and it then lies dormant in the uterus, in what is known as "fetal diapause." This is similar to the delayed implantation of higher mammals, except the fertilized egg is not implanted in the uterus wall. The egg remains in this state until the current joey has vacated the pouch. In most macropods, therefore (the eastern gray kangaroo is one known exception), there is an almost continual cycle of reproduction, with the newly emerged young being immediately replaced with a new embryo. A female kangaroo can have a large ex-pouch baby alongside her but still suckling, a baby in the pouch, and a blastocyst in the uterus awaiting its turn, so she can therefore produce three babies in a period of just over two years. When the latest joey enters the pouch and attaches itself to a teat, the "used" teat—that is quite elongated by then, as it enlarges to fit the growing baby's mouth—is still available for the older joey, which pushes its head into the pouch to feed.

At birth marsupial young are helpless, naked, and unable to thermoregulate, and most of their fetal development then occurs in the pouch. When they emerge, about four months after entering the pouch by the baby opossums, eight months in the red kangaroo and 10 months in the gray kangaroo, they are furred and mobile and have reached the stage that coincides with the precocial placental mammals at birth. The tammar wallaby, one of the most well-studied species, cannot thermoregulate until is about six months old; this coincides with the functioning of its thyroid gland, whose hormones control metabolism—the conversion of oxygen and calories to energy.

Athough marsupials are synonymous with pouches, they are not present in all species. In fact they are totally absent in some and just folds of skin around the teats in others. Many of the South American opossums, including the mouse opossums (*Marmosa*) and shrew opossums (*Caenolestes*), lack pouches. In the marsupial mice (*Antechinus*), the pouch is absent but develops for the breeding season; the kultarr (*Antechinomis laniger*), for example, a species the size of a large mouse, grows folds of skin to cover its teats. The marsupial anteater or numbat (*Myrmecobius fasciatus*) also lacks a pouch. In the absence of a pouch, baby marsupials just dangle from a teat that swells in their mouth to firmly anchor them.

Pouches are well developed in the kangaroos, wallabies, koala, wombats, phalangers, and bandicoots. They open forward in most species, but to the rear in the koala, wombats, yapock, and the bandicoots. The pouch is most obvious when occupied by a large baby in the late stages of its development, especially when its arms and legs protrude. Only females have pouches, except for the aquatic yapock (*Chironectes minimus*), in which both sexes have them. The male's does not close tightly, but the female's is waterproof and she can submerge without risk to the young. The pouch provides similar protection for the embryo or fetus as the uterus does for a placental mammal embryo. Warmth, humidity, and food are all available to the naked and helpless offspring. The babies that grow in an enclosed pouch survive even though they are re-breathing air that has almost 20 times more carbon dioxide than oxygen.

Pouch-cleaning usually indicates that birth is imminent, and in preparation the female seeks a secluded place where she is unlikely to be disturbed, as this can prove fatal to the young. In all species, those with pouches into which the young must find their way, and the pouchless forms in which the young must quickly locate a teat and become attached, birth is a vulnerable time for the tiny naked and blind babies that cannot provide their own warmth. Despite appearing helpless, however, they make their own way to the pouch, or to a teat in the pouchless species, where they complete their development. Young marsupials do not have an active immune system at birth, and it does not function fully until just before they leave the pouch. Abandoned or orphaned babies are therefore very susceptible to infection. The young are also unable to control their body temperature (to thermoregulate) until they are well furred. The tammar wallaby does not begin to thermoregulate until it is almost six months old.

Marsupials with a forward-opening belly pouch sit in a manner that allows the newborn young the easiest access to the pouch. In the kangaroos, in which birthing has been filmed on many occasions, the female sits with her tail forward between her legs. The young have sharp claws and crawl up into the pouch without assistance, a journey that may take an hour. The mother's licking of the fur between the cloaca and pouch, originally believed to assist the young in their journey, is now thought to be merely cleaning up after the event. If the mother is forced to move during this crucial time, her baby will most likely fall to the ground, and she makes no attempt to retrieve it.

All the kangaroos, wallabies, and wombats have a single baby. Twins have been reported occasionally in all species, and although both babies may survive while attached to the teats (for up to three months), one then succumbs to competition for pouch space. The other species mainly have multiple young. The Tasmanian devil and the bandicoots have four babies, tiger cats have six, and the mouse opossums have up to 10 in each litter.

The number of teats has a great influence on marsupial reproduction, for young marsupials do not share the teats like kittens or piglets. Each baby must attach to a teat or die, whereas those of multi-offspring Eutherian mammals can jostle for the available teats, and it is possible for them all to survive if the mother has sufficient milk, and on occasion for more to survive than there are actual teats. An unusual reproductive feature of some nonmacropod marsupials is the incidence of superfetation, in which many more young are born than can possibly be raised because each baby needs its own teat. The quoll may give birth to 18 young, yet has only eight teats; and in the common opossum, the birth of up to 50 babies has been reported, yet the mother only has 13 teats, a seemingly excessive "safety measure" to ensure each teat gets a baby. Ensuring each teat is occupied appears to occur only in association with superfetation. The bandicoots for example, have eight mammae, or nipples, but rarely give birth to more than four young; and the mouse opossum has up to 13 mammae, yet it seldom has more than six babies. In the pouched species, the teats are always inside the pouches.

The newborn marsupial's lips are fused along their sides, leaving just a terminal mouth opening. Having reached a teat, it closes its mouth over it, and the teat then swells to fit the ridges of the youngster's hard palate. There are separate passages from the nasal cavity to both the esophagus and the trachea, so that it can suckle and breathe at the same time when attached to the teat. Female macropods have

Gray Kangaroos   *At this stage of its life, probably about nine months old, the baby kangaroo, or joey, fills the pouch and has probably already been leaving it for a couple of months for short periods, to rest alongside its mother and to graze. It vacates the pouch completely at the age of 10 months.*
Photo: Sandra May Caldwell, Shutterstock.com

four teats, but only the one to which a joey is attached actually lactates, the others apparently regressing until the next birth occurs. Most terrestrial species stay in their mother's pouch until they are weaned, but arboreal ones generally leave the pouch quite early and cling to their mother. The baby koala leaves the pouch when six months old and is carried by its mother. Teats cannot support the weight of the growing babies for long, and those that dangle from teats release them quite early. Baby mouse opossums release their mother's teats when about three weeks old and climb onto her back, where they stay until they are weaned at two months of age—one of the earliest weaning ages of all the marsupials. The other major type of baby care occurs in species that protect their young in a den or nest from the time

Gray Kangaroos *Now too large to enter the pouch, its mother allows the joey to suckle for a further four months. During this time, however, she will likely have her latest baby attached to another teat, the two functioning teats providing milk of different composition in keeping with the joey's age and requirements.*
Photo: Clive Roots

they leave the pouch until they are weaned. Examples are the Tasmanian devil babies that leave the pouch at four months, and are left in the den while their mother goes off to feed; and the sugar glider, in which the young leave the pouch when two months old but remain in the nest for another two months.

For years it was believed that a baby marsupial removed from the teat early in its development could neither be replaced nor given to another mother; but this untrue, at least intraspecifically. Pouch young of both red kangaroos and gray kangaroos have been transferred without difficulty or affecting their development to mothers of the same species, from which young of similar age had been removed. The transfers were made by simply placing the young in the pouch and allowing them to find the teat themselves. Transfers of pouch babies between mothers of different species have not been as successful. Very small joeys have also been returned to the pouches of anaesthetized kangaroos; and mothers that kept ejecting their babies have had their pouch opening sealed with surgical tape until the baby reattached itself to a teat.

Placental mammal milk has a relatively stable composition, and although changes do occur during the lactation period, they are minor compared to the dramatic ones that occur in marsupial milk composition during the growth of the babies. Initially marsupial milk has a high sugar content and is low in fat, but when the joey is almost ready to leave the pouch for the first time, the milk's energy content rises to almost four times that of the first week, as a result of a great increase in fat to about 60 percent of the milk solids. When the joey begins to grow hair, there is also an increase in the sulfur-containing amino acids. Changing the milk composition in this manner aids the development of the young, unlike the placental mammals, for which the major development occurs before birth and there is less reliance upon complex milk variations.

To accommodate the arrival of a new baby before the previous joey is weaned, which occurs in the macropods with their continuous breeding cycle, the mother supplies milk with a different composition to each teat. The teat to which the new embryo is attached provides a weak, sugary milk, while the older joey receives the enriched version. The traditional belief that the marsupial mother pumps her milk into the newly attached baby's mouth is untrue; the babies suck their milk in the normal manner.

Excluding the milk of the tammar wallaby, which contains galactose, marsupial milk has only traces of lactose, a disaccharide made up of glucose and galactose, which is the main carbohydrate in the milk of the placental mammals. Consequently the marsupial's evolutionary development did not include production of the enzyme lactase, which breaks down lactose. Baby kangaroos hand-raised with cow's milk have shown great intolerance to it; the fermentation of the undigested sugars in their intestines fosters bacterial growth and gastrointestinal disturbances. Also, cataracts have developed due to the subsequent buildup of the sugar galactose, which was converted into alcohol and deposited in the lens. However, recent studies in Australia indicate that marsupials may differ in their tolerance to lactose, and this tolerance may also vary according to the joey's age. Marsupials also lack the enzyme sucrase and therefore cannot digest sucrose. Consequently, lactose- and sucrose-free milks are used to hand-raise joeys, and specialized milk replacers have been produced commercially for the marsupials, with several grades of milk for kangaroos, wombats, and possums to match the natural changes in the composition of their mothers' milk. There is also a colostrum replacer for orphaned marsupials.

Marsupial colostrum, a clear liquid produced as the first milk, contains the immunoglobulins or antibodies that provide protection against infection. Unlike newborn ruminants that benefit from colostrum for no more than 36 hours, the baby kangaroo's intestine can absorb parental colostral immunoglobulins for up to 200 days, which corresponds with the period spent in the pouch. A gastric groove, similar to that of young ruminants to direct the milk past the rumen into the acidic stomach, is also present in the marsupials. However, it is believed to be of value only when they are being weaned, as the newborn marsupial's stomach is not differentiated and can digest milk, unlike the compartmentalized stomach of the ungulates. When babies leave the pouch and begin to eat herbage but are still nursing, the groove would be of value to direct milk straight into the acidic stomach, bypassing the rumen where the herbage is being fermented.

The time young marsupials spend in the pouch varies according to the species, and although there is a great deal of specific variation in the raising process, four basic stages are involved. These are detachment from the teat, the first appearance of the head, complete emergence but still returning, and then the final, permanent emergence. Even when they are too large to reenter the pouch, they still insert their head to suckle. The red kangaroo joey detaches from the teat when ten weeks old and is said to be "free-living" in the pouch. Its head appears at five months, and it emerges at eight months but continues to suckle until it is one year old. Tree kangaroo joeys spend more time in the pouch than all other marsupials. Their heads appear and they begin nibbling vegetation when they are eight months old; they leave the pouch to feed at 10 months, and finally vacate it when they are 14 months old.

■ SOME OF THE SPECIES

Yapock (*Chironectes minimus*)

Also known as the water opossum, the yapock is the only semiaquatic marsupial; it lives on the banks of rivers and lakes in Central and tropical South America. It seeks all its food in the water, eating mainly crustaceans (shrimps and crayfish), molluscs (snails and shellfish), aquatic insects, and frog spawn; but although it is a good swimmer, it rarely catches fish. Its adaptations for a lifestyle spent mostly in water, when not sleeping, are webbed hind feet, a streamlined body, short and dense water-repellent fur, and small ears that fold over when underwater. It has long, coarse facial bristles, like the otters, for locating prey, but also uses its tactile forepaws, holding them out in front in murky water. Its forefeet are unwebbed; it lacks the opposable thumb of the arboreal opossums, and it has pads on its fingertips. The yapock has a head and body length of up to 16 inches (40 cm), and its tail is the same length, bare and rat-like, furred only at the base and with a yellowish-white tip. Although prehensile, it is used only for gathering nesting material, carried curled up against its belly. Its coat is marbled gray and black, and its belly is yellowish white. The yapock's nest is usually in a burrow, with an entry hole above the water line, but it also builds nests in the undergrowth on the water's edge. It is nocturnal and sleeps all day in its burrow, so it is rarely seen and little is known of its lifestyle. Both sexes have a pouch, but the opening to the males' does not close tightly. The female yapock gives birth to six young, which she keeps in her rear-opening pouch initially even when she enters the water, as a well-developed sphincter muscle closes the opening. When the babies are too large to fit comfortably in the pouch, she leaves them in the nest.

Bare-tailed Woolly Opossum (*Caluromys philander*)

This is one of the normally pouchless New World opossums. But it develops one when it has babies to carry, although it is rather rudimentary and not like the deep pockets of the kangaroos. The regular litter is four young, usually born twice annually; the babies attach themselves to the mother's teats and are carried for three to four months. They then spend another month with her in a tree nest. There are three species of woolly opossums in the genus *Caluromys,* and this one lives in South America's primary and secondary rain forest, east of the Andes and from the Caribbean coast south to southern Brazil. It has a soft, thick coat, reddish-brown above, with dark brown eye rings and a yellowish-gray belly. Its head is gray and bears three facial stripes, one running down the bridge of the muzzle from the crown to the nose, and one from each eye ring to the nose. The woolly opossum's head and body reach 11 inches (27 cm) in length. It has a very long tail, at least 15 inches (38 cm) long and furred only for the first quarter of its length, with the balance being bare and rat-like, but strong and prehensile. The woolly opossum has a pointed nose, forward-facing eyes, and naked ears. It is arboreal and nocturnal and climbs about in the treetops after dark, searching for fruit and insects. Like many small mammals it is short-lived, three years being considered a good life span. Its gestation period is 24 days, and the newborn young are ⅜ inch (10 mm) long and weigh 0.0045 ounces (0.13 g). They are solitary animals, socializing only at mating time, and are not believed to establish a territory, living instead a nomadic lifestyle.

Tammar Wallaby (*Macropus eugenii*)

This is the smallest wallaby, with a head and body length of 18 inches (46 cm), a 12-inch-long (30 cm) tail, and weighing about 15 pounds (7 kg). It originally ranged across southwestern Australia, where it lived in scrub and semidesert habitat, but was exterminated by 1920 as a result of hunting, loss of habitat, and predation by introduced foxes. It has since been reintroduced, with individuals from the population that has thrived on New Zealand's Kawau Island since they were themselves introduced there late in the nineteenth century by Governor Sir George Grey. Dark grayish brown with buffy-brown underparts and reddish legs and arms, the tammar wallaby was the first kangaroo seen by westerners—in 1629 by Dutch sailors stranded in the Wallaby Islands—long before Captain Cook's more famous first sighting of kangaroos at Botany Bay. It is nocturnal and shelters during the day in dense vegetation, appearing at dusk to graze in grassy areas.

The tammar wallaby is a well-studied species, especially regarding the composition of its milk. Its gestation period is 28 days, and the milk is rich in carbohydrates for the first seven months, but then the carbohydrates drop and the protein and fat increase quite dramatically. The joey detaches from the teat when it is about three months old but stays in the pouch, suckling ad lib for the next three months. Its eyes open and it grows a coat of fur between four and five months of age, and it begins to regulate its own temperature when it is six months old, when its thyroid begins to function, and it then leaves the pouch for the first time. It vacates the pouch permanently when eight months old, although it may continue to suckle until it is 11 months old. In addition to the change in composition, this wallaby practices synchronous lactation, in which two mammary glands produce different milks—providing one teat with milk for the older joey, and another with milk of a different composition for the latest baby that has just entered the pouch, allowing the nursing of two offspring with greatly different nutritional requirements. The other change in tammar milk composition within the lactation period is in the milk sugars, from lactose in the first week only to mainly galactose thereafter, so bottle-raised orphans have been unable to cope with the high lactose content of cow's milk.

Wallaroo (*Macropus robustus*)

The wallaroo is a large and powerfully built kangaroo, midway in size between the larger wallabies and the largest species—the red and great gray kangaroos. Males weigh up to 120 pounds (54 kg) and are shaggy-coated, dark gray animals that become almost black with age and have a bare black snout. Females are considerably smaller and are much paler than the males.

Wallaroos are rock-dwellers, agile climbers with furry pads on their feet for traction, that live in the rocky and mountainous regions of Australia, excluding Tasmania. A subspecies known as the euro lives in Western Australia. The wallaroo is adapted for survival in a very hot and dry climate, where little standing water is available. It is nocturnal and rarely moves far from the shelter of the rocks, avoiding the heat of the day by sleeping in hollows among the boulders or under rock overhangs. It even digs holes in the soil to rest in, and only ventures out at dusk to graze on the flat meadows between the outcrops. It can survive without free water if green vegetation is available, and digs deeper holes, down 39 inches (1 m), to find water. It pants heavily to induce evaporative cooling, and concentrates its urine to conserve water. The wallaroo is mainly a solitary animal, but may live in small family groups of a female and her offspring from the previous two or three years. After a gestation period of 31 days, the joey wallaroo is attached to the teat for almost three months; it sticks its head out of the pouch for the first time at five months, and it is expelled at the age of nine months.

Numbat or Marsupial Anteater (*Myrmecobius fasciatus*)

The numbat is a marsupial that filled a niche in its new land and developed into the equivalent of the anteaters of its ancestral New World homeland. It is a classic example of convergence—the evolution of similar characteristics in unrelated animals as a result of adopting a similar lifestyle. It has a typical anteater's long and sticky tongue, extending about 4 inches (10 cm) from its mouth, and it lives mainly on termites and ants collected from rotten wood and dead trees, and located with its acute sense of smell. It often climbs high into trees searching for termites' nests. It has small, weak teeth, but it chews up soldier termites, and swallows others whole. It has even developed a similar type of bony palate to the pangolins or scaly anteaters of Africa and Asia.

The numbat has a laterally flattened head and body, about 9 inches (23 cm) long and a bushy tail that is 7 inches (17 cm) long. It has a pointed head and ears and large eyes, and its main body color is brick red, with white and dark alternating bars across the back from the shoulders to the rump and a black stripe bordered with white from the ears to the nose, through the eyes. The tail is often fluffed out like a bottle brush. It is so unique it is the only member of the family *Myrmecobiidae*. A diurnal animal, unlike most of its marsupial relatives, the numbat may dig a short nest tunnel, but prefers to sleep in hollow logs at night, in a nest of leaves and grass. It lives in open scrub woodland in southwestern Western Australia, where it lives a solitary existence. Numbats lack pouches, and their babies attach themselves to one of the four teats and are dragged along under their mother's belly. They remain attached for four months and then move to the mother's nest for two months.

Ring-tailed Possum (*Pseudochirus peregrinus*)

A common woodland species of eastern Australia, Tasmania, and southern Western Australia, the ring-tailed possum has invaded suburban gardens and houses, where it makes its home in roofs and attics. Quiet and retiring animals, they are social and usually live in groups of a mature male with one or two females and their young of the previous year. The ring-tail has short grayish-brown fur with a pale belly, and its babies are reddish brown. It reaches a head and body length of about 14 inches (35 cm) and has a long and tapering, bare prehensile tail of a similar length that is curled into a ring and held against the side of the body when not in use. It has dense, soft, and woolly fur, grayish brown above with yellowish white underparts, and it weighs up to 34 ounces (1 kg) when adult. The ring-tail is nocturnal and sleeps by day in a nest of leaves and grasses in the undergrowth or in a hollow tree, to which it carries nesting material in its curled tail. It has a well-developed pouch that opens to the rear and contains four teats, although it rarely raises more than two young. They leave the pouch when they are four months old and are then carried on the mother's back for at least two more months. The father may also carry the young, a rare example of paternal care in the marsupials. The ring-tail is a folivore, an animal that eats mainly leaves and flowers and supplements with fruit and nectar; and like the koala, it can digest eucalypt leaves. Although it is quite safe in the trees, except perhaps from the nocturnal native cats, on the ground it is preyed upon by foxes and dingoes.

Marsupial Mole (*Notoryctes typhlops*)

The most unusual marsupial, and the only member of its genus, this mole is a highly specialized animal adapted for a subterranean existence and consequently evolved to resemble the moles of the rest of the world, a case of parallel evolution in the marsupials and eutherian mammals. Australia's native mole is about 6 inches (15 cm) long and has a thick silky coat that varies in color from white to golden red depending upon the substrate it occupies. It is superbly adapted for a fossorial or burrowing existence, with a powerful and rigid body—as its neck vertebrae are fused—and with a horny shield over its nose, a short leathery tail, and a very sensitive snout. Its claws are enlarged for digging, especially the third and fourth claws

Koala   *Despite its tree-living lifestyle—mostly in a vertical position—the mother koala has a backwards-opening pouch. Its baby is fully furred when six months old and is presumably able to thermoregulate then, for it leaves the pouch and clings to its mother. It also begins the weaning process then, suckling and eating semidigested eucalypt leaves from its mother's cloaca.*
Photo: John Austin, Shutterstock.com

on the forefeet, and those on the three middle digits of the hind toes. The horny nose shield is used as a bore, pushed forward by its strong feet, allowing the loose sand to fall in behind it. It is blind, its tiny eyes lacking a pupil, lens, and optic nerve. It is also assumed to have poor hearing, as it lacks external ears and has tiny ear openings, but is probably sensitive to vibrations. Little is known of its underground behavior, but breeding is believed to occur in firmer sand where a permanent burrow can be made, and it carries its single baby in a pouch. The marsupial mole lives in sandy desert regions in the western half of Australia, and although fossorial, it frequently comes to the surface. Its diet is believed to be primarily earthworms and beetle and moth larvae.

Kowari or Byrne's Marsupial Mouse (*Dasyuroides byrnei*)

A rat-like marsupial, with a head and body length of 6 inches (15 cm) and a tail of 4½ inches (11 cm), the kowari has a coat of dense and soft grayish fur with creamy underparts. The terminal half of its tail is heavily tufted with bushy black hair. It is a polyprotodont or carnivorous marsupial, differing from the herbivorous species by its multiple incisor teeth and its lack of syndactylous or joined toes. It lives in the desert regions of central Australia, in southwestern Queensland and neighboring South Australia and the Northern Territory; but it is becoming increasingly rare due to habitat loss from sheep grazing, and is now on Australia's vulnerable species list. The kowari has a poorly developed pouch, little more than two folds of skin on either side of the six nipples. Its gestation period averages 33 days and it gives birth to five young, usually twice annually. It is a colonial species, an unusual social arrangement for carnivorous marsupials although the males are aggressive to each other, and lives in extensive interconnected burrow systems, in which nests of grass are made. Kowaris spend the hot daylight hours in their burrows and come out at night to search for insects, small birds, lizards, and rodents. They get sufficient moisture from their prey and do not need to drink.

8 Den Mothers

This chapter is the first of four on the eutherian or placental mammals, whose young are more advanced at birth than the monotremes and marsupials. There is no taxonomic relevance in grouping them in this manner, and the categories are based purely on parental behavior. The only relationship between the animals within this chapter, and each of the following three chapters, is the condition of their young at birth. The greater development of the placental mammals at birth results from a longer gestation period made possible by the possession of a placenta, the organ from which the higher mammals take their name and the reason why they are so successful. It is richly supplied with blood vessels and acts as a barrier between the mother's system and the developing embryo. Consequently, white blood cells and other immune system components are kept within the boundaries of their own systems while allowing nutrients (sugars, proteins, fats, and minerals) to pass in, and waste products to pass out, of the embryo's environment. This clever system allows young mammals to spend more time in their mother's womb, and in some cases they are born in such an advanced state, they can run with their mother just a few hours after birth and are said to be precocial. But many are helpless and little more advanced than young marsupials at birth, and are therefore altricial.

Like the marsupial's pouch babies and the monotreme's hatchlings, altricial placental babies cannot regulate their body temperature and need complete protection from the elements as well as from predators. They must be kept warm and dry and are therefore raised in a nest, burrow, or den, the warmth being provided by the nest material and close contact with their siblings, but mostly by their mothers. Mammals that bear altricial young and care for them in seclusion include the carnivores (excluding the hyenas), most rodents, rabbits, some primates, and several insectivores such as the shrews and hedgehogs. The most unusual members of this group are the pigs—the only ungulates, or hoofed mammals, to give birth to altricial young, and the only ones to have litters. Altricial babies are helpless at birth;

their ears and eyes are closed, and they may be naked like most rodents, rabbits, and hedgehogs; sparsely haired like black bear and grizzly bear cubs; or well furred like kittens and puppies. They are immobile; their movements are confined to wriggling about in the nest or finding a teat. A characteristic of these mammal species is the production of large litters, such as the dozen babies produced by the house mouse and black rat, compared to the one or two infants of precocial rodents such as the guinea pig and the paca.

In addition to differing from all other placental mammals in the state of their young at birth, these altricial species are the only mammals in which there is considerable pre-birth care. Before giving birth, they must dig maternity burrows, make nests, or find and prepare dens. The many mammals included in the following chapters—those who carry their babies, whose young follow them, or whose pups are simply left on a beach—make little or no external arrangements for the birth of their young.

Their mother's milk is not enough for newborn altricial babies. Helpless and unable to thermoregulate initially, their survival demands warm and dry conditions. Unlike precocial seal pups that are well endowed with fat at birth, altricial young have little or no body fat, and all species therefore spend their first weeks, and in some cases months, in a nest, den, or burrow. They may dig their own burrow like the European rabbit (*Oryctolagus cuniculus*) or use the burrows of others like the warthog, who usurps aardvark or crested porcupine homes. Cottontails line their ground nest with their own fur to keep their naked babies warm and cover the

House Mouse Pups   *Newborn house mice are naked, blind, helpless, and unable to maintain their own body temperature. They are typical altricial young, which need the warmth and security of a den or nest. Mouse mothers are very caring parents, to the extreme of killing their young if they are concerned for their safety.*

Photo: Max Blain, Shutterstock.com

nest with grass when they leave; and a vole's nest, resembling a ball of short pieces of dry grass, may be protected beneath an old piece of plastic or metal in a hedgerow. Arboreal nesters like the squirrels, black rat, and some dormice may use an old bird's nest or the holes drilled by woodpeckers as the base for their own home, or they may make a large leafy nest like the gray squirrel and the ruffed lemur. Rodents living in close association with humans may build their nest in the loft or attic of a house or barn, or in sewers or culverts. The wild cats and dogs, whose newborn young are furred, are usually born directly onto the floor of a cave or burrow and depend upon their siblings and mother for warmth. Bear cubs, born like polar bears in a snow den, or black bears in a hole beneath tree roots, are kept warm by their mothers curling around them as they suckle.

Endocrine changes occuring a short time before parturition prompt seclusion, finding a den, building a nest, or whatever is typical for the species, and prepare the female for the physical arrival of the young. There is a drop in the levels of progesterone (that has held lactation in check during the gestation period) and an increase in the hormones estrogen (which stimulates the development of secondary sexual characteristics) and prolactin (the pituitary gland hormone that stimulates breast development and the production of milk). After birth, the level of estrogen drops in most species, except those that mate again soon after birth, but prolactin stays high and progesterone levels remain low. Although endocrine production initiates preparedness for birth, the continuance of care after birth (or postpartum, as this stage is called) depends on the sensory stimulation (olfactory, visual, and audio) provided by the young. Surprisingly, prolactin levels also increase in male rodents that help to care for their young, and this increase coincides with a decrease in testosterone; these hormonal changes result from stimuli received from either their mate or from the offspring.

In the seasonally cold regions of the world—the temperate and alpine regions —the timing of the birth of altricial babies is critical, and it must coincide with the most favorable time of year for raising the young. Even the hibernating bears achieve this, through a combination of delayed implantation of the fertilized egg in the wall of the uterus, and birth and raising in the hibernation den to avoid winter; the cubs then appearing furred and mobile in spring. Births occur throughout the day and night, but especially at night, when mothers are less likely to be disturbed and have the cover of darkness. Delivery time varies considerably, but is usually lengthy in the altricial species because they generally have large litters. A wild boar sow may take several hours to deliver 12 piglets, a time of great anxiety for the mother, and any stress may delay the process even further.

Despite their dedicated post-birth care, the parents of altricial babies do not initially have well-developed infant-recognition mechanisms, and it may therefore be possible to introduce other babies from closely related species—from a jungle cat to a domesticated cat for example. The mothering instinct is so strong in some parents they may accept unrelated babies, like the lactating dog that raises lion cubs, or the tiger that nurses piglets, but this occurs only in artificial situations. The parents of precocial young have stronger infant-mother bonding mechanisms, and a farmer may have to drape the skin of a dead lamb over an orphaned one before the mother who lost her baby will foster the orphan.

■ PARENTAL CARE

Although instances of foster care occur occasionally in the wild, it is generally true to say that baby animals cannot survive without their mother's care, at least initially when they need to suckle. Thereafter, in some species the fathers also help to raise their young, and in a few species other helpers are also involved. Help depends largely upon the animal's mating system. Polygyny, in which the males mate with several females, is the most common system, and results in maternal care. Only about 3 percent of all mammals live in monogamous pairs; this provides opportunities for biparental care, as the males can be certain the babies they are helping to raise are actually theirs. Most species in which paternal care is highly developed have altricial babies, when there is greater advantage for the female to have help raising the young. A father's duties include defending the territory, guarding the babies, bringing food for them, grooming and playing with them, and in some species (in the following chapter), carrying them. When males help to raise the young, the mother is free to mate again, or to conserve her energy and be better prepared for the production of her next litter.

In addition to paternal care, there are examples in the placental mammals of other family members helping to raise the young. In social species such as the banded mongoose (*Mungos mungo*), in which only the dominant female breeds, the other females get firsthand experience through witnessing the birth and then helping to raise the babies, thus improving their own parenting abilities when they eventually get the opportunity to breed and establish their own group. Another form of assistance, called cooperative polyandry, occurs in the African wild dog (*Lycaon pictus*), in which the social group includes several mature males. In a mating system that contradicts the normal rule, in which males attempt to improve their chances of reproduction by mating with more than one female, two or more wild dogs may mate with a female and then help her raise the offspring, which they may not have sired. In an even more unusual form of cooperative care, the lifestyle of the naked mole rat (*Heterocephalus glaber*) resembles that of the honey bee, with a breeding "queen," consorts, and workers who care for the young and forego the chance to breed for the benefit of the colony.

Although there are many conflicting accounts on the mating systems of the wild cats, polygyny seems to be the most common form of breeding. In the solitary cats—all species except the lion—males generally mate with more than one female during the breeding season, despite some very credible reports of a male Bengal tiger cohabiting with just one female during the breeding season, and not even being distracted by other females in heat in the area. The case for the European wild cat is equally confusing; a pair has been seen in a monogamous arrangement, with the male even helping to raise the kittens; whereas females normally mate with several males, with resulting multiple paternity in the same litter, followed by purely maternal care. Some biologists therefore consider monogamy in many mammals to be a myth. However, polygamous behavior does not imply life together for lengthy periods, except for lions, the only social cats, that live in packs called "prides." These are usually composed of several related females, with one or more resident males who fight for leadership and must occasionally battle outside coalitions of males attempting to depose them. The females usually breed synchro-nously, and then raise their cubs communally, the cubs suckling from any female. Whatever their mating system, however, it is rare for pairs of cats to remain together until the young are born, and even rarer for males to assist in raising the babies. Cheetahs are solitary animals, and meet only for mating. The mother is totally responsible for raising her cubs, and gives birth in the undergrowth, but they are very vulnerable when she is away hunting, and she moves them frequently to reduce the chances of other predators finding them. They begin to follow her at the age of six weeks, and stay with her, learning how to hunt, until they are 18 months old.

Unlike all other carnivores, the hyenas have precocial young, but their babies remain in their dens for some time after birth. Brown hyenas (*Hyena brunnea*) live in clans in a large communal den, but females give birth in smaller satellite dens from which they have usurped aardvarks or wart hogs. The cubs stay in the den for three months, appearing at the entrance only when their mother calls them out to nurse. They then move to the communal den, from where they emerge more frequently until they finally go foraging when they are one year old. The spotted hyena (*Crocuta crocuta*) has a more promiscuous mating system, as the large groups of females cannot be controlled by a single male and therefore associate with several males. The pups are raised by their mother alone, but first they must successfully pass the "sibling challenge." They are well developed at birth, with a coat of black fur, their eyes open, and teeth that have already erupted. Although they cannot walk, they pull themselves along with their front paws. Same-sex siblings fight for dominance within hours of their birth, and these fights are often fatal. When the surviving pups reach the age of about three weeks, their mother takes them from their birth den to the clan den, where they mingle with other pups. Accounts of their nursing vary, however, from mothers nursing only their own pups, to a com-munal arrangement of pups suckling from any lactating female.

Several wild dogs are monogamous and form long-term pair bonds, in which the father assists in raising the young. The raccoon dog (*Nyctereuctes procyonoides*), the only canid to hibernate, and the maned wolf (*Chrysocyon brachyurus*) are monogamous and have partners for life, although the latter is normally solitary out-side the breeding season. In both species, the fathers help to raise the young. Pater-nal care also occurs in the jackals and the bat-eared fox (*Otocyon megalotis*). Despite living in large packs, the social wild dogs are also normally monogamous. Bush dogs (*Speothos venaticus*) form long-term pair bonds and live in packs with their off-spring of various ages; but only the dominant pair breeds, and the whole pack then helps the parents raise the young. In the most familiar social hunter, the timber wolf (*Canis lupus*), packs normally comprise related animals, including a mated pair and their offspring of various ages, with the alpha male acting as pack leader, although this animal may not be the mate of the breeding female. However, group size is con-trolled by the fact that only the dominant female breeds and other pack members assist the pair in caring for the young and look after them when the pack goes hunt-ing. The hunters bring food back to the den and regurgitate it for the pups.

Pigs are the only ungulates to build nests and have litters of young. Their young vary in body covering at birth. Those of the bearded pig (*Sus barbatus*) are quite well haired, and wild boar (*Sus scrofa*) piglets are sparsely haired but soon grow their striped coats;, whereas newborn warthogs (*Phacochoerus africanus*) and babirusa (*Babyrousa babyrussa*) young are virtually hairless, although neither species has much hair even as adults. Their piglets lack body fat and must be kept warm in a nest, especially until they can thermoregulate, which is unknown for wild piglets, but begins at the age of four or five days in the domesticated pig. Warthogs live in family groups, in which the females generally have a lasting bond with each other, whereas males are solitary and polygynous. They are the only pigs to typically have their young in a burrow, usually one borrowed from an aardvark, although they may dig their own. Warthog sows isolate themselves to farrow and provide sole care for their piglets, who may remain in the burrow for six weeks. The other pigs all nest above ground. The bush pig (*Potamochaerus porcus*) makes a nest of dry grass resembling a miniature haystack, in which her piglets lie quiet and still when she is away. The bush pig is socially monogamous, and lives in herds or "sounders" made up of a dominant boar and the sow, plus their young of all ages; and both parents stand guard over the nest. In contrast, the male wild boar (*Sus scrofa*) is usually a solitary and polygynous animal, while several sows live in sounders with their offspring. The boars visit only for mating and leave the sows to raise their litters unaided.

Insectivores are typically altricial. Shrews are born naked, blind, and deaf, in a nest of grass and leaves in a protected place, often in a tree stump. Baby moles are similarly helpless at birth, but are always born underground in a nest within a tunnel. An unusual variation of male care is practiced by the tupaia or tree shrew (*Tupaia glis*), which lives in monogamous pairs. After helping the female make a special maternity nest, the male continues to live with her in their regular nest, while she visits the maternity nest only every two days to feed the babies.

All bears are polygynous. The males are loners who mate the females within their territory. The females are then left to make their own arrangements for the birth, which involves finding a den—in the case of the northern species, a den suitable for the winter hibernation and raising the young during the long sleep. Sow bears raise their cubs without help from the fathers. All bears have altricial young, born in a very immature state. Giant panda and polar bear cubs are covered with fine, short hair at birth; whereas the other species are very thinly haired or hairless, and very tiny compared to their mothers. When it is about one month old, the giant panda takes her cub with her when she goes out to find bamboo, but the other northern bears stay in the shelter of the den until spring.

The members of the great family *Muridae*, containing over 1,100 species of mice, rats, gerbils, hamsters, voles, mole rats, and other rodents, have, with few exceptions,[1] fully altricial babies at birth. They are blind, helpless, and naked, and would chill quickly unless nursed. Most species are polygnandrous or promiscuous, in which both males and females may have several partners, and the females raise their young without assistance. They include the familiar vermin species—the brown rat, black rat, and house mouse—plus the meadow vole, montane vole, grasshopper mouse, and most hamsters. In some hamsters, the golden hamster and common hamster for example, the females are extremely aggressive, solitary animals; males enter their territories only for breeding and are driven out afterward.

Polar Bears *A typical den mother, the polar bear gives birth and suckles her cub in the safety of a den dug into a snowbank. Venturing outside when the cub is about three months old, she must be wary of all male polar bears, even her cub's father, as they consider all cubs fair game. Cubs stay with their mothers for at least two years, learning how to hunt seals.*
Photo: Courtesy USFWS, photo by Scott Schliebe

In total contrast, dwarf hamsters are monogamous; and in at least the Dzungarian or Campbell's hamster (*Phodopus campbelli*), the male assists in caring for the young, and even cleans them as his mate gives birth. Expectant mothers of the other dwarf hamsters kept as pets, the Siberian hamster and Roborovski hamster, are less tolerant of their mates at birth, but welcome them back soon afterward. Other monogamous species in which care is biparental include the prairie vole, pine vole, and the deer mouse. The male deer mouse guards the young when the mother goes off to eat. He covers them with bedding and may even wash them, improving their chances of survival, which is exactly why both parents care for their young.

Like the hamsters, mole rats also differ completely in their parental care. The naked mole rat is a communal or eusocial species in which many animals live together in a group, ruled by a queen. In contrast, the blind mole rat (*Nannospalax*) of the Mediterranean regions has evolved a solitary lifestyle and lives alone in its burrow. The pregnant female makes a big mound above ground, like a large mole hill, in which she builds a nest for her litter of four babies, which are born naked and helpless.

Rabbits, which are not rodents but close relatives, also have altricial young, born underground in a nest in the "warren"; whereas cottontail babies are similarly naked and helpless but are born on the ground in a nest of fur pulled from the mother's body. In contrast, hares and jack rabbit babies are fully furred and have their eyes are open at birth.

The babies of most primates cling tightly to their mothers and are then carried everywhere; but the young of many prosimians, or primitive primates, cannot hang on, so their mothers make nests for them.[2] Ruffed lemurs (*Varecia variegata*) build a nest of leafy branches; they are often said to be the only primates to build a nest, but this is untrue. The dwarf lemurs and the bushbabies also build large nests of branches and vines in the treetops, and several other lemurs build nests in tree cavities. Another unique aspect of their reproduction is the birth of multiple young, the red ruffed lemur having up to six young, and the fat-tailed dwarf lemur four babies. At birth, their babies are furred and their eyes are partially open, but they are helpless and cannot cling to their mothers. Fathers are not involved in the care of young bushbabies, and when the nocturnal mothers leave the nest at dusk they carry their babies (usually one, but occasionally two) in their mouth and leave them on a branch to be collected when they return to the nest at dawn. Ruffed lemur mothers leave their young in the nest until they are mobile at two weeks old, but they change nests frequently and then carry their babies in their mouths to the new nest. Maternal care is practiced by polygynous prosimians, such as Coquerel's dwarf lemur (*Mirza coquereli*) and the russet mouse lemur (*Microcebus rufus*), both solitary species that meet just for mating; whereas the more social gray mouse lemur females live in groups and nest communally also, although they only care for their own young in the nest. Biparental care occurs in monogamous species such as the ruffed lemur and the fat-tailed dwarf lemur (*Cheirogaleus major*), which have permanent pair bonds, and the fathers help the mothers to raise the young. On occasions when a fat-tailed dwarf lemur male was unable to help his mate raise the family, the young all died.

The actual composition of milk varies in the altricial nesting and denning species. Cat's milk has 10.9 percent fat and 11.1 percent protein, whereas dog's milk has 8.3 percent fat and 9.5 percent protein. The milk of the domesticated rat is very high in fat (almost 15 percent) and protein (11 percent), while the bear's milk contains 30 percent fat. The milk composition changes, although not dramatically, during the lactation period. During the 45-day nursing period of the dog, for example, the protein and fat increase in the middle of the period, then drop back, and there is a large increase in the calcium content. In the altricial species that produce large litters, there are usually sufficient teats or nipples for all the young to share. But the milk supply to individual nipples is known to vary, and siblings compete for the most productive ones. In cats, the rear nipples produce the most milk; and in pigs, the front mammary glands are most productive. Consequently, the growth rates of the young vary according to their success in competing for the best milk supply, and the larger young at birth therefore have an advantage over their smaller siblings.

The period of intense parental care—the suckling period from birth to weaning —eventually comes to an end, its duration being loosely related to the animals' size and whether it is domesticated or wild. The end of parental care in domesticated animals generally coincides with weaning, simply because the human "parent" then takes over, providing weaning pellets for piglets, growers rations for chickens, or chow for the puppy. Wild animal babies stay with their parents while there are still survival benefits to be gained, and for them weaning may not be synonymous with independence, for it depends largely upon the amount of training needed to prepare the young for adulthood. This takes longer in the predators than in prey species.

The process of weaning the young off milk is prompted by the babies increasing demands for food as they grow, coupled with a decline in the mother's milk production, forcing the young to seek other foods. During this period, some parents bring solid food back to the den. Termination is not just in response to nutritional changes, however, but also hinges on the benefits of learned behavior that will increase the offspring's survival prospects, and is always related to the cost to the mother versus the benefits to the young. In the hunters, the young of tiny species like the least weasel go hunting with their mother for several weeks before setting off on their own; whereas tiger cubs stay with their mother for at least two years learning how to survive. Post-weaning parental care involves training, as the behavior of red fox parents illustrates. The father brings food to the vixen and teaches the pups survival skills, including how to bury food and then to locate it, "ambushing" them, and teaching them how to defend themselves. In contrast, the young of the almost continually breeding rats and mice must disperse soon after being weaned to make room for the next litter. Hunting demands greater skills than does simply searching for grain or cropping grass.

Despite their mothers' tender, loving care and a warm and cozy nest, altricial babies face many dangers, and reaching independence is not guaranteed. The danger comes not only from the many predators that have their own young to feed, but also from others closer to home. Siblicide—the killing of nest mates—which occurs in several birds, is rare in mammals and is unknown in altricial species. Infanticide, however, when babies are killed by their parents or other adults, is a much greater danger to young and helpless mammals, and in the altricial species it occurs in cats, dogs, and rodents. The major threat to lion cubs, for example, comes not from hyenas or leopards but from other lions. There is bitter, intense rivalry between prides, and the lions (both male and female) take every opportunity to kill the cubs of other prides. Also, when a new male takes over a pride, he kills all the cubs under one year old to delete the genes of his predecessor, and to make the females receptive sooner so he can begin his own family. Older cubs may escape, and occasionally females leave the pride with their cubs to save them, or they may even gang up against a new male to protect their cubs. New males also kill cubs born shortly after they have taken over a pride, as they know they did not sire them; but females may preempt this behavior by aborting their pregnancy.

Many rodents are also baby killers. Male lemmings, golden hamsters, house mice, and brown rats kill the offspring of others, and infanticide and cannibalism has been induced in laboratory rodents through malnutrition, even just a deficiency of vitamins. When rat and mouse mothers are stressed, or when they consider the environment too hostile to raise babies, they will eat them to recover some of their "investment"—the energy expended during the embryo's development in the uterus, and then the production of milk. On other occasions, they may practice "sexually selective infanticide," which occurs when stressed female polygamous rodents kill their male babies. This is because the opportunities for breeding are different

African Lions *A quiet moment for a lioness and her cub. Her "den" was probably a patch of dense bush, where she sought seclusion from the pride to give birth. Lion cubs are furred, but are blind and helpless at birth, and when they can walk at six weeks old they are taken back to the pride, where they may also nurse from other lactating females.*
Photo: smeyf, Shutterstock.com

for the sexes in polygamous species, in which the males or females may have several partners. A female reaching maturity is almost certain to breed, whereas a male must become socially dominant to do so. Even poorly developed daughters produced in times of stress are therefore still likely to breed, whereas weakling males will never become dominant, so raising them would be a waste of energy. This provides yet another example, although a very harsh one, of the major purpose of parental care—to increase the fitness and therefore the survival chances of the offspring.

Many ground squirrels practice infanticide. It has been reported in the California ground squirrel, Columbia ground squirrel, the yellow-bellied marmot, and the black-tailed prairie dog. Despite the prairie dogs' early warning system, when sentinels sit on the mounds and bark an alarm when a threat is spotted, the greatest danger to the infants actually comes from their own kind. Nonparental infanticide is the major cause of death of young prairie dogs, resulting in the loss of total litters and accounting for a large percentage of the annual crop. Surprisingly, it is mothers who enter the burrows of related females and kill their babies, possibly to cannibalize them and thus gain nutrients, or perhaps to ensure the continuance of their own line, but the real reason is unknown.

■ SOME OF THE SPECIES

Multimammate Mouse (*Mastomys coucha*)

This mouse, a little larger than a house mouse, is one of the most productive rodents. It has 12 pairs of teats or mammae, more than any other mammal except the tailless tenrec, and 22 embryos have been taken from laboratory animals, although the average litter is 10 babies. The multimammate mouse has a head and body length of 5 inches (12 cm) and a tail of similar length, and weighs about 3.5 ounces (100 g). It varies in color from light to dark buffy yellow, with grayish underparts, and its feet are white on top. It is mainly nocturnal and lives among rocks, tree roots, in tall grass, and in deep leaf litter; but it usually gives birth in a nest of grass in a burrow, of its own digging or borrowed from another rodent. It can produce a litter every 33 days.

The multimammate mouse is one of the most common rodents in Africa, distributed from the Mediterranean coast to Cape Province, in savannah, scrub, and open woodland, but it avoids the wet forests and deserts. It eats seeds, fruit, and insects and is a pest in croplands. It is also the host of the lassa virus, which is often fatal to humans. In the wild, the mouse is polygynous; the males mate with several females, each one then raising her young without assistance. Captive mice have been kept in pairs, however, and the male was tolerated by the female and even assisted at the birth, helping to clean the babies and then eating the placenta. The pups are born naked or very sparsely haired, blind, and helpless; they stay in the nest for 10 days and are weaned when 20 days old.

Campbell's Dwarf Hamster (*Phodopus campbelli*)

Unlike the common pet hamster—the Syrian hamster, a very solitary animal in which the aggressive females only tolerate males briefly at mating time, and are sometimes not too sociable to people—the dwarf hamster is a peaceful and tolerant monogamous species, and care of the young is biparental. Dwarf hamsters live contentedly in pairs, and the male is allowed to be present at birth; it even helps to clean the pups, afterward staying near when the mother nurses them. The male's estrogen and cortisol levels rise prior to his mate giving birth, then drop afterward as testosterone rises. Dwarf hamsters are charming animals, with soft pale brown fur on their backs, black ears, and a black dorsal stripe. The sides of the face, lower flanks, limbs, tiny furred tail and hairy feet (they are also known as hairy-footed hamsters) are white. They have cheek pouches, which they fill with food and then push out with the forepaws into their stockpile. Like many small rodents, they are very fecund and can breed when just seven weeks old. The average litter is eight pups, born after a gestation period of 18 days, and the mother can be mated again four days after parturition. The pups are born naked and blind and are helpless, although they can pull themselves around with their forelimbs. But they grow rapidly; their skin darkens when they are just three days old, and they are furred by the age of two weeks. They gnaw on solids when eight days old, before their eyes have fully opened, and they are weaned at three weeks, when their mother is already preparing for the birth of her next litter. Campbell's dwarf hamsters are natives of the arid plains and sand dunes of central Asia—northwestern Mongolia and Kazakstan—and are also often called Russian or Dzungarian hamsters, after the Mongolian kingdom of that name that existed in the eighteenth century. They are now prominent in the pet trade, but to add to the confusion created by so many names, two other species, the Siberian or "winter white" hamster (*Phodopus sungorus*), which becomes white in cold weather, and the Roborovski dwarf hamster (*Phodopus roborovskii*), which lacks the dorsal strip, are also offered as pets. Also, they are often collectively referred to as Russian hamsters or just dwarf hamsters.

African Crested Porcupine (*Hystrix cristata*)

This is the largest and most impressive porcupine, its body covered with long, hairy spines and sharp quills, and with a large and "blunt" head in which the eyes are set back on its sides. It is a native of Africa outside the equatorial forest zone, Italy, Sicily and Asia Minor. A related species occurs in southern Asia, and at the end of its range in the Caucasus, a subspecies called the hairy-nosed porcupine (*H. indicus hirsutirostris*) is the northernmost porcupine. An adult crested porcupine can reach a length of 32 inches (80 cm) and have a tail 5 inches (15 cm) long with open-ended quills, and a mature male may weigh 66 pounds (30 kg). The quills lie flat against the body, but when used for defense they are raised and the porcupine rushes backwards. Prior to this, as a warning, it rattles them, when some may fall out, which gave rise to the fallacy that it can shoot its quills. Crested porcupines are nocturnal and hide during the day in caves, in wide rock crevices, or in burrows that they dig themselves or have been vacated by aardvarks. These large rodents are vegetarians and cause great damage to crops, as they are wasteful feeders. They are also known to eat carrion, and they gnaw on bones to obtain calcium.

Crested porcupines live in groups comprising the monogamous pair and their large offspring. They breed throughout the year, but females prefer to give birth in a separate den and raise their young there for the first two weeks, when they return to the communal den and care then becomes biparental. Litters contain up to four young, born after a gestation period of 115 days. The babies are altricial, but are covered with bristles and soft quills that harden in a few days. Although the babies may leave the den for the first time when they are 10 days old, eat solids then, and survive without milk if necessary, they are usually not weaned until they are three months old.

Naked Mole Rat (*Heterocephalus glaber*)

The smallest of the mole rats and the only member of its genus, the naked mole rat has wrinkled yellowish skin and is virtually naked except for a few long straw-colored hairs, especially on its feet. It is only about 3½ inches (9 cm) long; has a 1½-inch (4 cm) tail, and weighs 1.4 ounces (40 g). The mole rat has a short head, tiny ears, tiny eyes (it is virtually blind), very long and protruding incisor teeth, and large forefeet with flattened toes and small claws. It is a completely fossorial or burrowing animal of the deserts and semideserts of Ethiopia, Somalia, and Kenya. A highly social species, it lives in groups of up to 100 individuals in large and complex burrow systems, with a communal nest chamber and foraging tunnels that may extend just below the surface for up to 650 feet (200 m). Mole rats dig with their teeth and push the loosened soil back with their feet and up into mounds on the surface. They have no subcutaneous fat, and their skin lacks sweat glands, so they have very poor thermoregulation, with a low body temperature of about 89.6° F (32°C). The temperature in their burrows is similar to their body temperature, although the humidity is quite high—usually about 90 percent—and they avoid tunnels near the surface during the heat of the day.

Mole rats are the only eusocial mammals—species that care for their young co-operatively—and their society is a matriarchal one. The dominant animal is the only breeding female, called a queen, who inhibits reproduction in the others. Like a queen bee, she spends most of her time in the nest chamber in the company of a few nonworking individuals, and they are brought food by the worker mole rats. Although the workers are fertile, they do not breed; several dominant males are selected as the queen's consorts, and she produces an average of 10 young per litter, several times each year. The queen nurses the babies for about one month, and their rearing is completed by the workers, who feed them feces to wean them onto the roots and tubers of their adult diet. The workers protect the colony, using their large incisors against the snakes that enter their tunnels and are their major preda-tors. When weaned, the babies start working in the colony, although they are not fully mature until they are almost one year old.

Least Weasel (*Mustela nivalis*)

This is the smallest carnivore, a hyperactive animal with a very high metabolism that needs 40 percent of its body weight in food daily. A fast runner, it feeds mainly on mice, voles, moles, birds, and insects, killing its vertebrate prey with a bite at the base of the neck. When it is able to kill excessively, it stores food in its den. The least weasel has a slim and cylindrical body, about 8 inches (20 cm) long, with a 1-inch-long (2.5 cm) tail, and weighs just 3 ounces (85 g). It is chestnut brown above and has white underparts, and white feet, but it is very secretive and is rarely seen. It is a native of the Palaearctic Region—the northern parts of North America and Eurasia, including North Africa—although it is absent from Ireland and the Arabian Peninsula; and it has been introduced into New Zealand. It is at home almost everywhere there is unspoiled vegetation and food, in grassland, woodland, scrubland, and mountainsides, where it is quite territorial and marks the bounda-ries of its land with excretions of its anal glands. In the northern, colder parts of its range, it becomes white in winter like the larger stoat or ermine, but it lacks that animal's black-tipped tail. The least weasel is a totally solitary species except at mat-ing time, and it lives mainly in abandoned or usurped rodent or rabbit burrows, where the naked, blind, and helpless young (up to six) are born in a nest of grass. The mother provides all the care, and the young are weaned at four to five weeks, but stay with her learning to hunt until they are 10 weeks old.

Bengal Tiger (*Panthera t. tigris*)

One of the largest cats (another subspecies, the Siberian tiger, is larger), the Bengal tiger is generally considered a polygynous animal, in which there is no permanent pair bond and a male may mate with several females, staying with each one for several weeks only during the mating season. Consequently, parental care is provided entirely by the mother. This tiger lives in India, Bangladesh, Myanmar, Bhutan, and Nepal, in forest, dry woodland, grassland, and rocky Himalayan foothill regions, where it occupies a territory of perhaps 20 square miles (51 sq km). A very powerful killer, an adult male can weigh up to 570 pounds (260 kg), but females are slightly smaller. With its huge canine teeth, up to 4 inches (10 cm) long, it can overcome the largest wild cattle, such as the gaur and water buffalo, plus chital, nilgai, and wild pig, and it occasionally kills elephant calves. It is claimed that an adult tiger can eat up to 88 pounds (40 kg) of flesh per day, covering any leftovers with grass and leaves for use over a number of days. Of all the tigers, the Bengal tiger has most frequently become a man-killer, particularly in India. It is a stalk-and-ambush hunter, because it cannot chase prey for long distances, and kills by biting into the neck and severing the spinal cord, or by holding the preys' muzzle and suffocating it, while clinging onto the animal's back.

All subspecies of the tiger are now rare or virtually extinct; the Bengal tiger is the most plentiful, with perhaps 4,000 animals barely surviving in the face of hunting and loss of habitat. It is only relatively safe in the national parks and is most plentiful in the vast mangrove swamps of the Sundarbans, in eastern India and Bangladesh. Tigers usually have three or perhaps four cubs, but it is rare for more than two to be raised, usually due to the difficulty of finding enough food. Cubs are furred but helpless at birth, which usually occurs in a cave, a large hollow tree, or in dense vegetation. They stay with their mother for two years learning how to hunt and survive.

Spotted Hyena (*Crocuta crocuta*)

The largest and most powerful of the three hyenas, the spotted hyena weighs up to 175 pounds (80 kg) and has a shoulder height of 36 inches (90 cm). The characteristic hyena shape of large and powerful head, neck, and shoulders sloping down to the weak hindquarters, is most obvious in this species; and it has a short tufted tail and shorter ears than the other hyenas. The spotted hyena is yellowish gray with dark spots over the whole body and upper legs. It has well-developed sight, hearing, and smell; and in proportion to its size, its jaws are the most powerful of all animals, capable of crushing the largest bones to reach their marrow. An efficient hunter, it gets more of its food from killing than scavenging, and it is now believed that lions more often take advantage of hyena kills than the reverse. It is very fast over short distances, reaching a speed of 36 miles per hour (60 kph) and able to run down wildebeest, large antelope, and zebra, particularly individuals weakened by age or sickness.

The spotted hyena's original, historical range, Africa south of the Sahara, has been reduced in the last century by loss of habitat and through persecution as a predator of livestock and threat to sleeping humans in tents and native huts. They are now rare in many parts of their former range, particularly in southern Africa. They are nocturnal and den by day in burrows of their own digging or those usurped from porcupines, aardvarks, and warthogs. Some individuals are solitary,

Spotted Hyenas   *The spotted hyena is the only carnivore to have precocial young that are well furred, have their eyes open, and are partially mobile when born. She gives birth in a special and separate maternity den, and when the pups are about three weeks old she carries them to the communal den, where they grow up in the company of the clan adults and the other females' pups.*
Photo: SF Photography, Shutterstock.com

but most are highly social and promiscuous. They live in large clans divided into small hunting packs in which a female is dominant, but there is apparently no permanent pair bonding within the packs, and females also mate wandering males who are not members of their own clan. Their social life is centered around a communal den, but the young are usually born in a separate satellite den some distance from the main one.

Unlike the other species included in this chapter, hyena cubs are not altricial, despite being raised in a den. After a gestation period of 105 days, the cubs are born fully developed, with their eyes open, and with fully formed teeth. Within hours of their birth, same-sex siblings battle for dominance, biting and shaking each other. The dominant one then prevents the other from nursing, and it soon dies. This rivalry is believed to kill one-quarter of all spotted hyena cubs, and is the only known case of siblicide in mammals. There is no rivalry between cubs of the opposite sex. About three weeks after their birth, the mother carries them to the communal den, where they mingle with the clan, but she continues to nurse them. Her milk is probably the richest of any terrestrial carnivore, containing almost 15 percent protein and 14 percent fat, and this sustains the cubs when she is away hunting for several days at a time. Cubs are not weaned until they are one year old.

Giant Panda (*Ailuropoda melanoleuca*)

The giant panda is the most familiar bear, with a white body and black hind-limbs and forelimbs, a black band over the shoulders, black ears, and black eye patches. When adult it has a shoulder height of 33 inches (85 cm) and weighs 350 pounds (160 kg). It is modified for gathering and eating bamboo, and an unusual modification of its forepaws allows the first digit to be flexed against the pad for grasping stalks. It has a much larger head for its size than the other bears, plus powerful jaw muscles, heavy, broad molars, and a leathery gullet. Yet despite these adaptations, it has not evolved a digestive system that can cope with the highly fibrous nature of its diet. The giant panda is the rarest bear, with perhaps 1,000 animals surviving in the wild, in fragmented but protected, areas in central China. There are less than 100 captive animals, mostly in China, where the first captive birth occurred at the Beijing Zoo in 1963. After a slow start, breeding results are beginning to improve, especially as a result of artificial insemination.

The panda lives in mountain bamboo forests, usually above 8,200 feet (2,500 m), but moves lower in winter. It was once considered a close relative of the raccoon, but DNA testing proved it is a bear. It is diurnal and crepuscular, but may stay up late in the evening, as it needs to feed continually for many hours each day to gain sufficient energy from its very poor diet. Care of the young is totally the mother's responsibility, but unlike all other northern bears she does not hibernate, simply because she cannot store adequate fat for a long sleep on a bamboo diet. So she shelters in a cave, a crevice, or a hollow tree, usually open to the elements; there she gives birth in late summer, usually to twins, although only one may be raised. Newborn cubs are blind, helpless, and toothless, and are sparsely covered with white fur. They are very vulnerable to dholes, lynx, yellow-throated martens, and foxes when she leaves them in the den during her long feeding sessions. When they are one month old—by which time they are fully furred—she begins to take them with her, clutched to her chest with one paw, and then caches them near the bamboo where she is feeding. They are weaned at the age of eight months, but stay with her for another year.

Asiatic Black Bear (*Ursus thibetanus*)

This bear is the Eurasian equivalent of North America's black bear, and in the northern parts of its range it must similarly hibernate to escape low temperatures and the lack of food. During this period the sow gives birth and raises her cubs in the den. Bears hibernate singly and care of the young is solely maternal. The Asiatic black bear is a very distinctive species, with a long, black coat and a white crescent-shaped patch on the chest. The fur around the throat and shoulders is longer and mane-like. Males weigh up to 440 pounds (200 kg), and females reach 275 pounds (125 kg). Its range is central and southern Asia, from Iran and Afghanistan eastwards to Malaya, Vietnam, and Taiwan, where its size and aggressive nature make it a serious adversary of humans, and it causes many deaths annually. It prefers forested regions, especially at higher elevations, but also occurs in the semiarid regions of southeastern Iran and neighboring Pakistan.

The Asiatic black bear is a typical omnivore, eating virtually anything of plant or animal origin and increasing its intake in the fall to store fat for the winter. While hibernating in a cave, in a hollow tree, or in a hole dug beneath the root mass of a fallen tree, the bear's temperature drops only to 88°F (31°C) from its normal 100° F (37.8°C). Its heart rate and breathing rate also drop, and its metabolism is therefore significantly reduced, resulting in a major saving of energy. This is not the profound torpidity of the ground squirrels, but the bear sleeps soundly, although it normally awakens quickly. The higher body temperature is necessary for the sleeping bear to give birth and to produce fat-rich milk to nurse its cubs and keep them warm, which a deeply torpid animal could not do. Its cubs, usually two, are born blind and virtually hairless and weigh 8 ounces (250 g). But on their mother's rich milk (containing 25 percent fat), they grow quickly; they are mobile and fully furred when they leave the den at the age of three months in March or April, when they weigh about 17 pounds (8 kg). They stay with their mother until the summer of the following year.

Dwarf Mongoose (*Helogale parva*)

This is the smallest mongoose, just 10 inches (25 cm) long, with an 8-inch (20 cm) tail, and a short but dense coat of speckled brownish gray. Like the other small species—the yellow mongoose, meerkat, and banded mongoose—it is a very gregarious animal, living in family bands of up to 30 individuals. The band usually comprises an old monogamous pair—of which the female is dominant—and their many offspring. All the other members of the group are socially subordinate and do not breed, but all help in raising the young. When the babies reach weaning age, the other members of the pack bring food, allowing the mother to conserve her energy. The group may disband when the matriarch dies. The dwarf mongoose lives in grasslands, open woodlands, and rocky scrub in the eastern half of Africa from Ethiopia to South Africa. It is terrestrial and diurnal, and dens in termite's mounds, rock crevices, and hollow trees, and occasionally in its own burrows. It is omnivorous and eats invertebrates and small vertebrates including rodents, snakes, lizards, and young birds, plus fruit. The female leads the band, while the male is alert for threats and intrusions into their territory by rival packs. Immigrant mongooses may join the band, and provide the same care as the family members. This cooperative care stems from lack of opportunity for offspring to acquire a habitat of their own, but nonbreeding females may shortcut their wait for the dominant female to die or stop breeding by transferring to another pack, where they may stand a better chance.

African Hunting Dog (*Lycaon pictus*)

The African hunting dog is a pack hunter—the wolf of sub-Saharan Africa—where packs have a very large territory and are always on the move, being almost nomadic. They live in the drier, open country (savannah and wooded grasslands), and their packs have a most unusual composition. The normal arrangement of a group is several adult males, a female, and her young and maturing female offspring, which move away when they are about 18 months old. Young males also break away from groups to form their own packs, and are then joined by a number of females from another pack. One of the females becomes dominant, and the others then move on to find another group of bachelor males, which is becoming increasingly difficult as their numbers dwindle. The only time hunting dogs stay anywhere for long is when they den, either in burrows of their own making or in those of aardvarks or warthogs, to give birth and raise their pups. African hunting dogs are one of the few mammals with blotched patterning, and no two animals are alike; their short and sparse coats are mottled with black, yellow, and white. They have long legs, large ears, and wide and powerful jaws. Males and females are similar in size, reaching a weight of 77 pounds (35 kg). They are most active at dusk and dawn, and, after locating their prey—antelope, zebra, or wildebeest—by sight, they approach as close as possible, then select an individual and pursue it at high speed for several miles, snapping at the animal's rump and flanks. There appears to be little ranking among the pack, although the same male may lead the chase each time. These dogs are now rare in many parts of their original range, due to loss of habitat, reduction of their prey, and persecution as a livestock predator.

Bearded Pig (*Sus barbatus*)

This is a large wild pig of the rain forest of Malaysia, Sumatra, and Borneo, where it is mostly active after dark. It has an unusual shape, with a very slender body, thin legs, and an elongated head, with large warts under each eye that are very protuberant in the males. Adult boars are about 66 inches (1.7 m) long, with a shoulder height of 32 inches (80 cm), and weigh 333 pounds (150 kg). Their coat is coarse and dark brown to gray, and they have a short tufted tail and thick, tufted yellowish whiskers on the sides of the face that meet over the bridge of the nose. Bearded pigs live in small herds and are believed to be polygynous, the males mating with several females in the group. But parental care is purely maternal, and the pregnant sow leaves the herd when she is about to give birth and finds a secluded dense patch of vegetation where she makes a large nest of plant material, which may be 78 inches (2 m) wide by 39 inches (1 m) high. Her piglets, usually three, are born in the center of this mound, and at birth they have a striped yellowish and dark brown coat. They remain hidden in the nest for at least two weeks, and then leave with their mother and rejoin the herd. They are weaned at three months of age, but stay with their mother until they are a year old. Bearded pigs eat fallen fruit, roots, shoots, invertebrates, and carrion; and they follow troops of monkeys for the fruit they drop from the treetops. They are a favorite food themselves of tigers, leopards, clouded leopards, large pythons, and humans. They are the only pigs to undertake annual migrations, congregating in herds of several hundred animals to follow traditional trails to a different area.

Ruffed Lemur (*Varecia variegata*)

There are two subspecies of the ruffed lemur, the black-and-white ruffed lemur (*Varecia v. variegata*), and the red ruffed lemur (*V. variegata ruber*). They both live in the eastern rain forests of Madagascar, but the red ruffed subspecies is restricted to the Masoala Peninsula of the northeast. They are large lemurs, with a head and body length of 20 inches (50 cm) and a slightly longer, thickly furred tail, and weigh 10 pounds (4.5 kg). The black-and-white ruffed lemur has a black-and-white body, black tail and face, and a white collar and white tufted ears. The red ruffed lemur has a reddish body; black tail, hands, feet, and face; and a white nape. Most primates carry their babies, but the babies of the ruffed lemurs, like those of several other prosimians or primitive primates, cannot cling onto their mothers, and therefore spend their early days or weeks in a nest. Female ruffed lemurs make a large nest of leafy branches and moss for their young, at least 50 feet (15 m) high in a rain forest tree. Also unique for primates, they have litters of young. Up to six offspring have been recorded for the red ruffed lemur, although three is more common; but there is a high mortality rate, due mainly to babies dying from falls. The young are well furred at birth, and their eyes are open, but they are helpless—the main criterion for altricial young—and remain in the nest for at least two weeks. The pregnant female makes a nest away from the group, but for some reason she often changes nests, carrying her babies in her mouth to the new one. She may also carry them to a branch, where she deposits them while she is foraging. Ruffed lemurs live in small groups of variable structure, sometimes a pair with older young or groups of multiple males and females, but the females are always dominant. The nest and its young are guarded by all members of the group including older siblings, which babysit the latest offspring and keep them warm, while the mother searches for fruit, blossoms, bird's eggs, and nestlings.

Notes

1. One is Australia's broad-toothed rat (*Mastacomys fuscus*), whose young are well developed and well furred at birth and attach themselves to a teat and are dragged around. Baby African swamp rats (*Otomys*) are also precocial at birth, but cling tightly to a teat for their first week. The young of the cane rat (*Thryonomys swinderianus*) are also furred, have their eyes open and are active soon after birth.

2. The baby fork-marked lemur (*Phaner furcifer*) is born in a tree-hole nest, but after a few days clings to its mother.

9 Baby Carriers

Most mammalian babies are just as clearly altricial or precocial as hatchling birds. The tiny, naked, and helpless newborn mouse is quite obviously altricial, while the well-haired bison calf that follows its mother soon after birth is certainly precocial. But there is another group of newborn mammals that does not fit either category. They do not meet all the criteria for altricial young, as they may be furred and their eyes may be open, yet others are naked and their eyes are closed; but they are all virtually helpless, so they are certainly not precocial. They are the babies that are carried by their mothers, to whom they cling with hands and feet and sometimes even their teeth.

The major groups of baby carriers are the bats and the primates, except for a few prosimians that provide a nest for their young. Anteaters, pangolins, sloths, and several rodents also carry their young. Only the primates have feet and hands suitable for holding their babies, which they do occasionally against their abdomens or chests; but they retain the skeletal structure of quadrupedal animals and usually walk or run on four feet, which is not conducive to carrying a baby in the hand or clutching it to the body. The mother gorilla is an exception, for she must carry her helpless baby until it is four months old, when it clings to her back. Otherwise, it is the babies' responsibility to hold on, and they get little or no assistance from their mothers. With few exceptions, the baby carriers do not make any pre-birth arrangements in the form of nest or den preparation. Their babies are born, hang on to their mother's coat, and she continues as normal. There are few exceptions to this rule, one being the terrestrial pangolins that give birth in a burrow, although this may not have been dug especially as a maternity ward, and they stay underground nursing their baby for one month before carrying it outside. The northern bats, and a few primates such as the Japanese macaque (*Macaca fuscata*), and the Hanuman langur (*Semnopithecus entellus*) and rhesus macaque (*Macaca mulatta*),

Yellow Baboons *Young baboons cling to their mother's belly for their first six weeks, then climb onto her back and later sit upright, riding jockey-style. They join play groups at the age of three months under the watchful eyes of their parents. They begin to eat solids at five months and are weaned at about eight months old, but depend upon their mother's guidance and protection until they are two years old.*
Photo: SF Photography, Shutterstock.com

whose ranges extend into mountainous regions of northern India and Pakistan, are the only temperate-climate mammals to carry their young; all other baby carriers live in warmer climes.

A few other animals also carry their babies for short periods, generally to provide a home for them. In the case of the giant panda this is temporary, for when her cub is furred, she carries it clutched to her chest from the den to the bamboo grove where she spends the day, caching it until she has finished foraging. The spotted hyena carries her young in her mouth from the maternity den to the communal den where they are raised.

■ PRIMATES

Most primates carry their babies. The higher or anthropoid primates carry them all the time, whereas the prosimians that carry their young (most lemurs and the lorises), as opposed to those that make nests (the bushbabies, ruffed lemur, mouse lemurs, and dwarf lemurs) care for their babies in one of two ways. The lemurs carry them; whereas the lorises "park" them on a branch when they leave at dusk to forage, then pick them up the next morning and nurse them. Most carrying species have single young. Twins are common in the Callithricids (the marmosets and tamarins), and they occasionally even have triplets; but as they only have two nipples, one of the three invariably dies. At first, baby primates are carried by their mothers, clinging ventrally where they are close to the nipples and are more protected; except young Callithricids, which cling to their mother's, father's, or sibling's sides and back. As they mature, young primates move around, and in the terrestrial species may eventually ride on their mothers' back, just above the rump, sitting upright like tiny jockeys.

Mothers are always the main, and usually the sole, carriers; but father Callithricids, titi monkeys, and douroucoulis take turns carrying their babies, and in the Callithricids large siblings also carry the latest offspring. In the two baby-carrying species that have been studied in detail, the common marmoset (*Callithrix jacchus*) and the cotton-top tamarin (*Saguinas oedipus*), prolactin levels increased and their level of progesterone decreased while they were carrying; these hormonal changes resulted from stimuli received from either their mate or from the offspring. Males also groom their young and guard them, but carrying the babies is considered the most valuable aspect of paternal care. Marmoset babies are very large for their mother's size, and their birth is obviously a demanding experience. The energy costs of raising two babies is very high; for in addition to their milk requirements, they must be carried at all times. The mother benefits from being able to rest, and having another male help the father carry them also relieves him of doing all the work.

Parental care in all the primates is a very lengthy business. Dwarf lemurs are dependent upon their mothers for at least four months. Young marmosets are independent at the age of five months, which coincides with the expected birth of the next litter. Infant dependency lengthens in the higher primates. Baby macaques and baboons are physically and emotionally dependent on their mothers for two years; and the baby chimpanzee is nursed and carried by its mother for four years, but is still dependent on her for another two years. Young of the Japanese macaque, which has the northernmost range of all primates, must contend with harsh winter weather when they are just a few months old, and they may soak with their mother in hot springs.

Polygyny, the form of polygamy in which a male mates with more than one female, is the most common breeding system in the primates. Males can generally defend and control a small group of females, whose long gestation periods and well-spaced receptiveness make them easier to control. These females then raise their young without assistance, other than the protection offered by the male. Typical polygynous species, in which a dominant male controls a harem of females, are lemurs, howler monkeys, capuchin monkeys, leaf monkeys, macaques, and the gorilla. However, some of these primates, plus the baboons, patas monkey, and the chimpanzee, are quite promiscuous. A receptive female Barbary ape may be mated by as many as 10 males. The pygmy chimpanzee, or bonobo, is considered the most promiscuous primate, in which several males mate indiscriminately with several females.

Polyandry, the other type of polygamy in which two males mate with the same female and assist in raising her young, is even rarer in primates but occurs in the marmosets and tamarins, which are the only higher primates to routinely have twins. There is just one breeding female and several males in most groups. This is unusual, because males generally need to make the most of their reproductive potential and mate several females. This arrangement does not appear to benefit the males, especially as one helps to raise the others' offspring, and the female does not increase her breeding potential by having two partners. Despite living in family groups, these small primates were originally thought to be monogamous, in the more typical family group arrangement in which the dominant pair monopolized the breeding. It is now known that the lone breeding female is mated by two males, which both share the care of the babies, particularly carrying them.

Monogamy, a bond between a single male and female, has evolved because a male could only defend one partner, probably because of the distance and availability of other females, or the inability of the niche to support larger groups. In view of the high investment by the male in his offspring, and almost certain paternity, males have a far more intensive care-giving relationship with their offspring. Monogamous species are the tarsier, titi monkeys, douroucouli, and the gibbons.

The composition of their milk is known for only a few species of primates, but certain assumptions are based on this limited knowledge. Primate's milk differs depending on which side of the Atlantic-Pacific Divide they live and whether they carry their babies all the time. Milk of the New World species (the *Cebidae*) that are all baby carriers is richer, having 4.6 percent protein, 5.4 percent fat, and about 6.8 percent lactose. The milk of the Old World primates varies. The ones known as "monkeys" (macaques, mangabeys, guenons. and baboons) of the family *Cercopithecidae*), also baby-carriers, have very dilute milk, containing only 1.7 percent protein, 4 percent fat, and 6.8 percent lactose. Milk of the great apes, which also carry their babies all the time, has slightly more protein but less fat and lactose than the other Old World species. The major difference in all primate milk composition, however, occurs in the prosimian species, whose milk is much richer in fat and protein. This is undoubtedly due to the fact that their babies are left unattended in a nest or on a branch for lengthy periods, whereas all other primate babies are carried all the time and have regular access to milk, which is consequently more dilute. Generally, as weaning begins, milk production declines, and its composition then increases in fat and protein.

Weaning time in the primates exemplifies the typical parent-offspring conflict, when the baby wants to suckle but the mother refuses to accept it until eventually

Spectacled Langurs *A mother spectacled langur, also known as the dusky leaf monkey, holds a protective arm around her baby. They are leaf-eaters, and to digest their highly fibrous diet they have a complex stomach similar to a cow, in which bacteria break down the cellulose. Spectacled langurs live in the forests of Southeast Asia.*
Photo: Clive Roots

the baby realizes its begging is fruitless and looks for its own food. Primates generally learn their social and sexual behavior and care of the young, by watching the activities of their parents or the other older members of the troop.

■ BATS

The bats vary considerably in their habits, and the only indisputable statement applicable to all species is that they can fly. Most bats carry their young, at least for part of their growing stages, but some of the colonial insectivorous species may leave their young in the birth cave. After hunting most of the night, they return only to nurse them; and perhaps never or rarely carry them outside. Care of the young is solely the mother's duty in at least 99 percent of bats, as few of the almost 1,000 known species of bats are monogamous. They include the African false vampire bat (*Cardiodema cor*), the South American false vampire bat (*Vampyrum spectrum*), and the yellow-winged bat (*Lavia frons*); and in keeping with the behavior of other monogamous mammals, they likely contribute to the care of their offspring by protecting them and perhaps bringing food back to the roost.

Most bats have a single baby, and a few have twins. Only the hairy-tailed bats (about eight species in the genus *Lasiurus*) occasionally have triplets. The pregnant bat's first duty is to find a safe place to give birth, and for many colonial species this means a traditional nursery roost separate from the colony's normal roost. These maternity roosts are generally warmer than their regular roosting places, the higher temperature assisting the development of their embryos and then later the growth of the young, as bats have difficulty thermoregulating and are somewhat dependant upon the ambient temperature.

Bats normally roost hanging head downwards, but most change position and hang by their thumbs, with their heads up, to give birth. Some form a pouch with their tail and interfemoral membrane to catch the baby. The young quickly seize a nipple, situated under the mother's armpits, and remain attached in flight. Twins are carried one on each side of her abdomen, to balance the weight, but a single baby usually hangs across her body, attached by its hind claws on one side in the armpit and stretched across her body to reach the nipple on the other side. The babies of many species are naked and blind, others are furred, and their eyes are open. The young, or "pups," of many colonial insectivorous species are left in a crèche soon after they are born, their mothers returning only to feed them, when they locate their own baby among thousands by their smell and calls.

The small insectivorous species are weaned and fly when just three weeks old, whereas the larger flying fox babies have a much lengthier growing period. They are carried by their mothers for their first four weeks until they are well furred, and are then left in tree roosts overnight. They can fly at 10 weeks old and are weaned at three months. The period of care for the vampire bat is almost eight months. Little is known about weaning in the bats and whether, for example, they carry food to their young, regurgitate food for them, or simply stop feeding them to force them to fly and seek their own. At least in the carnivorous false vampire bat, the father is known to bring food for the young; but alloparenting (when individuals other than the parents interact with the young) is only known in the vampire bat (*Desmodus rotundus*), in which adults feed orphaned bats in their group with regurgitated blood, and share their day's intake with others who did not find a victim the previous night. Even less is known about the composition of bats' milk. It appears to be quite rich in the insect-eating species, with a fat content of about 16 percent, 6.5 percent protein, and perhaps 2.5 percent lactose. In contrast, the fruit bats have a more dilute milk, with only 3 percent protein and 2 percent fat. It is unclear whether the northern bats continue to produce the same amount of milk for their young during their daily periods of torpor when they are roosting, when their body temperature drops to conserve energy, even in northern summers. Torpor may be suspended during the lactation period, or suckling may only occur when they are active at night.

■ OTHER CARRIERS

With one possible exception—the monogamous silky anteater (*Cyclopes didactylus*)—there is absolutely no paternal care in all the other mammals who carry their babies. The species that follow are all solitary, meet only for mating and then go their own way and the mothers raise their single baby without assistance. In an unusual behavioral coincidence both the "true" anteaters and the pangolins, or scaly anteaters, carry their offspring.

The giant anteater (*Myrmecophaga tridactyla*), a totally terrestrial species, and the arboreal tamadua (*Tamadua*) and silky anteater carry their young from birth until independence. These animals cannot provide solids for their young, so they must wean themselves by picking up their own ants and termites, not a difficult task as they are with their parent at the source of the food whenever she feeds. In the tamanduas, the most common anteaters, maternal care is extended and the baby stays with her mother for one year.

Baby pangolins (*Manidae*), in which there are both arboreal and terrestrial forms, ride low on their mother's backs, at the base of the wide tail. They are generally solitary animals, although a pair may share a burrow, but there are no records of males carrying babies. Like all the young that are carried on their mother's backs, baby pangolins must be sufficiently mobile and agile to move to her front to suckle; and when threatened, the mother wraps around her offspring as she curls into a ball. The African pangolin gives birth in her burrow and nurses the baby in her lap for the first month before venturing outside.

The three species of sloths all carry their single baby from birth, clinging tightly to their chest for about six months. The baby sloth begins to sample leaves at the age of three weeks, taking them from its mother's mouth and ingesting at the same time the specialized gut flora needed to digest them. Between three and four months of age it begins to pluck its own leaves.

The other mammals that carry their young do so in the most unusual manner. The babies attach themselves to their mother's nipples, like the young of the pouchless marsupials, and generally do not let go until they are weaned, so they are dragged around wherever she goes. They are all rodents, living in widely dispersed regions, so this strange behavior has developed independently, and in most cases their young are born fully furred. There is obviously no paternal care in these species, and the babies are assured a regular milk supply. In Africa, the swamp rats (*Otomys*) have up to four babies, fully furred and with their eyes open; and although mobile, they prefer to cling to a teat for their first week and are weaned when two weeks old. Bushrat (*Aethomys*) babies are dragged about attached to their mother's nipples for three weeks. The newly born pups of the South Africa's white-tailed rat (*Mystromys albicaudatus*) immediately cling to a nipple and are dragged around for almost three weeks. Australia also has several species of native rodents that "carry" their young in the same manner. The nocturnal tree rats (*Mesembriomys*)

Two-toed Sloth *The baby sloth is carried by its mother for up to nine months, holding firmly to the hair on her chest, as she hangs upside down in the forest canopy. It begins to take chewed leaves from her when about one month old, and a month later starts to pluck leaves within reach while still clinging tightly.*
Photo: SF Photography, Shutterstock.com

shelter in hollows during the day and forage at night with their babies attached, where they stay for two weeks. Baby Australian hopping or jerboa mice (*Notomys*) and broad-toothed rats (*Mastacomys*), also well furred at birth, are all dragged around attached to their mother's nipples. The North American deer mouse (*Peromyscus maniculatus*) has been observed carrying her babies for a short time in this manner. Many small rodents may "carry" their young attached to their nipples if they are disturbed while nursing and move quickly.

■ SOME OF THE SPECIES

Angwantibo (*Arctocebus calabarensis*)

The angwantibo, or golden potto, is a member of the prosimian family *Lorisidae,* and is therefore related to the Asiatic lorises, the African bushbabies, and to Bosman's potto. It is a unique animal, the only member of its genus, that lives in the equatorial rain forests of West Africa, specifically in Nigeria and Cameroon, where it is purely arboreal and nocturnal, preferring lower-level, dense vegetation. When adult, it is about 11 inches (28 cm) long, weighs 34 ounces (1 kg), and is virtually tailless. It has rounded ears and is stocky and bear-like, with yellowish-brown to fawn fur above, with paler underparts, and has a more pointed snout than the other lorises. A subspecies, the golden angwantibo (*A. c. aureus*), considered a full species by some zoologists, has bright golden-red fur. The angwantibo is a solitary animal that spends the daylight hours hiding in the foliage and, like its relatives, is very slow and methodical in its movements. Primarily an insect-eater, it apparently prefers caterpillars, but also eats fruit and is therefore considered an omnivore. It stalks living prey, moving slowly until it is within reach and then seizes them rapidly. Angwantibos are polygynous, the males having a large territory that overlaps those of several females, and they consort only at mating time. The females raise their single babies without help, and like the other lorises the baby clings to its mother's stomach; but as it grows, she leaves it on a branch while she is foraging. She still carries it after it is weaned at about four months old, but then it usually rides on her back. It is independent and leaves its mother when six months old.

Cotton-top Tamarin (*Saguinus oedipus*)

The tamarins and related marmosets are the smallest New World primates, monkey-like animals of the forests of tropical America. Externally, they are very similar, and their names are often used quite loosely and interchangeably. They differ mainly in the formation of their teeth—the marmosets have specialized incisors, with which they gouge holes in tree trunks to start the flow of sugary sap and gums, but the tamarin's teeth are not adapted for this purpose and they make use of the marmoset's gouges. The tamarins, and especially the pinche group from northwestern South America and Central America, to which the cotton-top tamarin belongs, have very sharp, tusk-like canines, and are more carnivorous, with a penchant for birds. They are strictly diurnal and spend the night in tree holes, of which they generally have several in their territory, and which they use alternately.

Like the other marmosets and tamarins, the cotton-top tamarin lives in extended family groups of about 10 animals, comprising a breeding pair and their offspring of varying ages. Only the dominant adult female breeds, and as the young females reach sexual maturity they are driven off to restrict the size of the group, or they may remain but do not breed. Twins are normal in this species, and a few days after birth they are carried by the males in the group and are handed back to the mother just for nursing. The cotton-top tamarin was originally believed to be monogamous, with just the dominant pair mating; but it is now clear that they are

Squirrel Monkeys *Too large to be carried ventrally, a squirrel monkey carries her big baby on her back. Aged about five months, it has already been leaving its mother for several weeks for short periods to explore and begin to eat solids, but is still reliant upon her. It may continue to suckle until it is over one year old.*
Photo: Simone van den Berg, Shutterstock.com

polyandrous, as two or more males mate with the dominant female and then help to raise her babies, a very rare mating system in primates.

Chimpanzee (*Pan troglodytes*)

The chimpanzee is still a fairly common animal in the great forest zone of west and central Africa. It is also the most common and certainly the most familiar great ape in zoological gardens. Smaller than the gorilla and orangutan, a wild adult male chimpanzee stands 63 inches (1.6 m) tall and weighs about 155 pounds (70 kg); females are slightly smaller. The species does not show the great sexual dimorphism of the other two great apes, but there is more variation in skin and hair color in chimpanzees, ranging from the typical black to pale brown. They have small heads, very large ears, and protrusive lips; and the face, ears, hands, and feet are black in adults, flesh colored in the young. Their arms are longer than their legs and reach below the knees, they have large hands with long fingers and short thumbs, and their big toes are opposable. Like the other great apes they are tailless.

Chimpanzees are good climbers but spend a lot of time on the ground foraging; they usually travel on the ground but seek the security of trees when threatened, and they make their sleeping nests there. They walk bipedally and quadrupedally with their arms straight and legs slightly bowed, and can move fast in this position. Chimpanzees are the most omnivorous of the great apes, although the bulk of their diet is fruit, leaves, blossoms, bark, honey, and invertebrates, especially termites. They also cooperate to hunt other vertebrates, including young baboons, bush pigs, and antelopes; and they sometimes cannibalize the young of other chimpanzee troops. They live in groups of up to 20 animals, and the females are generally promiscuous, although short-term pair bonds also occur. Their promiscuity is believed to be anti-infanticide behavior, as males uncertain of the paternity of the babies will not kill them, as they do the babies of unrelated males. Mother chimpanzees carry and nurse their babies for four to five years, and still protect them for two years afterward.

African Yellow-winged Bat (*Lavia frons*)

This bat is a small, bluish-gray animal with an olive-green rump and reddish-yellow wings and ears, weighs just over 1.2 ounces (32 g), and has a wingspan of 14 inches (36 cm). It lives in the savannahs and open woodlands of central Africa, from Gambia eastwards to Ethiopia and then south from there to Zambia. It prefers areas near rivers and lakes and is particularly associated with acacia trees, where it benefits from the insects attracted when the trees are flowering. Although nocturnal, it is often seen flying in bright sunlight. It is totally insectivorous, and has atypical bat hunting methods, hanging from a branch watching for insects to pass by and swooping out like a flycatcher to intercept them. It also has a most unusual social life, for it is territorial and, unlike most bats, monogamous. The bonded pair roost close together, and the male drives away other males that enter his small territory. One baby is born in spring just before the start of the rainy season and is carried by the mother until it is weaned at the age of 30 days, although she may leave it in the roost as weaning time approaches. But instead of clinging onto her fur in normal bat fashion, the baby yellow-winged bat actually clamps onto two false teats on the mother's lower abdomen and wraps its legs around her back. It stays with its parents for another month after it is weaned.

Rodriguez Fruit Bat (*Pteropus rodricensis*)

This is a large fruit bat found only on the island of Rodriguez, one of the Mascarene Islands in the Indian Ocean off the eastern coast of Madagascar, where it occurs in what is left of the dense rain forest. It is now very rare, having suffered from loss of habitat through deforestation, but its very low population of a few decades ago has been increased through captive breeding on the neighboring island of Mauritius and in zoological gardens worldwide, and its current population is about 4,000. The Rodriguez fruit bat is about 8 inches (20 cm) long, weighs 12 ounces (340 g), and has a wingspan of 35 inches (90 cm). It is variable in color—usually golden brown, but may also be orange, black, or reddish brown—and has a doglike face, large eyes, and widely spaced ears. It is a colonial species, but the sexes segregate themselves in the communal roosts. The single pup is born fully furred and with developed wings, and it clings to its mother's belly. She flies with it until it is 30 days old, when it has grown too heavy to carry; and is then left in the roost, where it clambers among the branches flapping its wings to strengthen them and developing its social skills with the other pups. It first flies when it is weaned at the age of two months, but continues to roost with its mother until it is fully independent at six months old.

Brown Long-eared Bat (*Plecotus auritus*)

A species in the family *Vespertilionidae,* a very large group of insectivorous bats with a worldwide distribution, the long-eared bat is a native of the Palaearctic Zoo-geographic Region, which stretches across Europe and Asia north of the Himalayas. It is now in decline due to the loss of habitat and roosting places, plus poisoning from hanging in buildings against wood treated with preservatives. It is a small bat, just 2 inches (5 cm) in head and body length, with a wingspan of 10 inches (25 cm), and is buffy brown with a pink face. It is totally insectivorous, and catches insects in flight, but can also hover like a hummingbird to pluck insects off leaves and buildings. The long-eared bat roosts in trees and buildings in summer, but uses caves in winter because of their more even temperature and humidity, which is necessary to prevent their wings from drying and cracking. A gregarious species, it lives in large colonies; and pregnant females move to caves used traditionally as maternity dens where the temperature is slightly higher, which aids the development of their young. Males may also join them there. A single baby is born, and although their mothers may carry them initially, they are normally left in crèches in the company of many other babies, their mothers returning to feed them. They can fly when three weeks old and are independent at the age of six weeks.

Giant Anteater (*Myrmecophaga tridactyla*)

The largest anteater and certainly the most impressive one, the giant anteater is a very distinctive animal with an extremely long, cylindrical snout, and a long and very bushy tail. Its head and body length can reach 42 inches (1.1 m), and its tail is 32 inches long (80 cm). It is dark grayish brown with a thin white line from the ear to the middle of the back, and has thick and powerful forelimbs. An adult male may weigh up to 110 pounds (50 kg), while the females are a little smaller. Its front toes are equipped with three long, sharp claws and a shorter one, and it walks on its knuckles with the claws curled inward to protect them. It prefers termites, and after tearing away the hard covering of the termitarium, its thin tongue—which can reach a length of 24 inches (60 cm) and is lubricated by the large salivary glands only when the animal is feeding—then penetrates deep into the passageways of the termite's nests.

The giant anteater lives in forests and grasslands from central America to Argentina. It is terrestrial but can climb into well-branched trees, and is a good swimmer. It sleeps in a secluded place with its bushy tail curled back over its body, and is mainly diurnal but often stays up very late, locating termites in the dark with its well-developed sense of smell. The giant anteater usually "gallops" rather clumsily away from danger, but protects itself when necessary with its claws, turning quickly and grabbing an aggressor with its feet and exerting tremendous pressure. The anteater's single baby is born fully furred and must make its own way through its mother's fur onto her back where it rides, but must regularly move around to her breast to nurse. Although it begins to run alongside her briefly when just two months old, it suckles for six months and is carried for a year.

Silky Anteater (*Cyclopes didactylus*)

This is the smallest anteater, a tiny animal only 6 inches (15 cm) in length, with a tail 8 inches (20 cm) long and weighing 8 ounces (225 g). Its range is the neotropical rain forest, from southern Mexico to Brazil and Bolivia, where it is arboreal and rarely comes down to the ground. It prefers the huge silk cotton or kapok tree (*Ceiba pentandra*) where its soft and silky coat of burnished gold is very cryptic, blending in with the sun-dappled foliage and the trees' large seed pods with their tassels of silk floss. It is purely nocturnal but needs to hide during the day from harpy eagles and crested eagles, while at night its main predators are boa constrictors and the spectacled owl. Although it is an inoffensive animal, it can inflict deep cuts with its sharp-clawed forefeet, which it raises in front of its face in a praying position and then jabs down swiftly. It has a prehensile tail—nature's finest tree-climbing aid—and extra-long soles that not only give a good grip on branches but are jointed and thus give a degree of opposability and an even better grip, so it is difficult to dislodge. It is believed to be monogamous, and the male is said to help in raising the single baby, and may also assist the female in feeding regurgitated termites to it during the weaning process.

Australian Hopping Mouse (*Notomys fuscus*)

This species of hopping mouse is one of several small, native rodents that are Australia's equivalent of the North American kangaroo rats. It lives in the drier regions of southern Australia, ranging north into southern Queensland, where it prefers sand dunes, semiarid grasslands and the desert scrub of the Outback. The hopping mouse reaches a length of 6½ inches (17 cm) and has a long mouselike tail, tufted at its end. It has long hind limbs and long and narrow hind feet, and it stands upright on its hind legs and bounds along like a kangaroo—a saltatorial mode of locomotion. It has soft, sandy-brown fur with a white belly, large bare ears, and large and strong incisor teeth. Hopping mice are mainly seed-eaters, and like the kangaroo rats rarely need to drink and have evolved a highly concentrated urine to excrete nitrogen without wasting water. Also, to combat dehydration, it is nocturnal and avoids the daytime heat in its burrows, which may be 48 inches (1.2 m) deep and are usually quite complex, with a nest chamber at the end. Hopping mice are believed to live in small groups, and to share their burrows with marsupial mice, but parental care is entirely the mother's responsibility. Four young are born in the nest after a gestation period of 35 days, but they immediately find a nipple and take hold of it, and are then dragged by the mother wherever she goes, until they are weaned when they are about one month old.

10 Mirror Images

In addition to the eutherian mammals that have altricial young and raise them in a den or nest, and those whose babies cling tightly to them, there are two other groups whose offspring are precocial. One of these is terrestrial and is the subject of this chapter. The other is semiaquatic and is discussed in the following chapter. Although precocial really describes the condition of the young at birth, it is also used generally to denote species whose young are well developed, and at birth the young of the precocial terrestrial mammals have a coat of hair[1] and their eyes are open. They are mobile soon after birth, but this does not necessarily mean they follow their parents immediately. Some do, but many remain hidden where they were born, for several days or even weeks. These babies are miniatures of their parents, resembling them in shape and usually in coat color and pattern, but differ in size. A major difference, however, is in their numbers, for the production of large, haired babies consumes considerably more of the mother's energy than producing a litter of tiny, naked young, and most precocial species have just one or two offspring after a long gestation period.

Precociality is not synonymous with independence, however; despite their advanced development, precocial young need parental care, usually for several weeks or even months, although a few are independent sooner. They all need milk, because baby mammals cannot survive without nursing, but parental care also includes protection from predators; their removal from potentially dangerous situations, and possibly physically defending them against some predators.

With few exceptions, the precocial mammals are prey animals, the herbivorous providers of protein to the carnivores, especially the larger species of cats and dogs, hyenas, and occasionally the terrestrial bears; and they must be continually aware of the danger. They are primarily the hoofed mammals or ungulates—the cloven-hoofed cattle, antelopes, deer, sheep, goats, peccaries, and similar animals; and

Burchell's Zebras *A perfect image of its parents, a Burchell's zebra foal stays close to its mother and the other herd members, for safety from lions and hyenas. Highly precocial, it can move with the herd within an hour of birth, and may nibble grass when only one week old. It is dependent on its mother's milk, however, for at least six months, and may continue to suckle until she foals again.*
Photo: Paul Vorwerk, Shutterstock.com

the solid-hoofed zebras and asses. In this large group of mammals, only the swine have altricial babies. The second major group of precocial species are the rodents, also major providers of protein to the world's carnivores, mostly the smaller species except in South America, where the capybara and spotted paca are favorite prey species of the two large cats—the jaguar and cougar. Most rodents have altricial young, but the members of the families *Caviidae* and *Octodontidae* (which includes the degu, a popular pet species) are precocial, as are several small rodents who "carry" their young attached to their nipples. The beaver, coypu, and the New World tree porcupines, and their close relatives the hares and jack rabbits, also have precocial young. Other mammals that give birth to precocial young are the two species of elephants, the elephant shrews and the hyraxes.

Precociality is synonymous with herbivory, and the only carnivores to have precocial young are the hyenas, whose pups are not independent killers until they are 18 months old. In contrast, the prey species must be able to escape quickly soon after birth or hide until they are more mobile. The social, herding mammals must keep on the move to find herbage, and their calves must keep up with them. Their advanced development at birth permits this, and even though they may continue to nurse for many months, perhaps until the birth of the next baby, learning to identify foods and potential predators is a much quicker process than that of the young lion or tiger learning to kill.

Pre-birth care is rare in the precocial species. It is lacking in the ungulates, other than finding a secluded place for parturition, and in some species a safe place for leaving the baby. The only precocial mammals in which parental care occurs prior to birth are the hares and jack rabbits, which make a nest in a shallow depression; and the burrowing rodents such as the mara and the paca, although they may use an existing burrow rather than make a new one. Beaver kits are always born in the lodge, which the parents provided for the family's winter security, but their mother does make a bed of wood chippings for them.

Most mammals, and therefore most precocial species, are polygynous, the mating system in which a dominant male mates the females in his territory. In most precocial species, this means his "harem" that he controls and protects, unlike the solitary carnivores such as the tiger and polar bear, which leave the females after mating. Polygynous species are mainly animals of the open range—the tundra, prairies, savannah, lightly treed grasslands, and mountain meadows. They include all the sheep and goats, most deer and antelope, and all the cattle except perhaps the anoas, the dwarf wild cattle of Sulawesi. They also include the zebras, wild horses, and asses, in which a stallion controls and defends a harem of several mares and their foals. The dominant male's care includes chasing off suitors, raising the alarm when a potential threat appears, and actively attempting to defend the herd against some predators. Although polygyny is usually associated with herds on the open range, two forest-dwelling antelope—the okapi and the bongo—are polygynous, but they are solitary animals, and the males have no association with the females after mating. There is also an extreme form of polygyny in the ungulates, known as the "lek system," in which a group of displaying males are visited by females just for mating. It is rarer in mammals than in birds, but does occur in the fallow deer, and in several African antelope—topi, Kafue lechwe, and the Uganda kob.

Sexual dimorphism, at least regarding size difference, varies in the polygynous species. It is especially pronounced in the musk ox, the African elephant, and the giraffe, where the males are considerably larger than the females. But in the zebras and asses, there is little size difference between the sexes. The major difference in most of these animals, however, lies in the possession of horns or antlers. In the deer family, *Cervidae,* antlers are borne only by the males except in the reindeer and caribou. In the cattle family *Bovidae,* males have horns, and many females also, although they are generally much smaller. Large body size, and the possession of larger antlers or horns than their conspecifics, confers an advantage during competition with other males to possess a harem. As a result of this mating system, however, when a dominant male controls numerous females, many males are surplus and never have the opportunity to pass on their genes.

Monogamy, in which a pair of animals remain together for a longer period than simply mating, sometimes for the season, the year, or even for life, is rare in mammals, and the few examples among the precocial species are all forest or bush animals. They include the South American forest deer called brockets (*Mazama*), and the huemal (*Hippocamelus bisulcus*), in which the pair remain together for most of the year, and the male challenges other males that attempt to enter his home range. Several dwarf antelope of the African rain forest and bush, such as the suni (*Neotragus moschatus*) and royal antelope (*Neotragus pygmaeus*), the klipspringer

(*Oreotragus oreotragus*), and the duikers, plus the mouse deer of Asia, are also monogamous. Monogamy is generally considered the mating system most likely to result in biparental care, as the male has a vested interest in caring for the young because he can be almost certain he is their father, and will thus increase their chances of survival and perpetuating his genes.

There is a limit to the care a male antelope can provide. He cannot carry his young like the primates, or bring food to mother and offspring like the wolf, or cuddle the baby like the male dwarf hamster. Nor can the mother, although raising the calf is entirely her responsibility, but in several species the fathers provide protection and support. The male klipspringer stands guard while his mate forages or nurses the calf, thus protecting his genetic investment. The male Kirk's dikdik (*Madoqua kirkii*) may groom his calf, and stays nearby while the baby is nursing, guarding against intruders and warning of potential danger.

■ RUNNERS

When they have recovered from the birthing process, precocial babies have two options—two primary survival strategies, as they are called—to run or to hide. Generally, it is the open-range species that run soon after birth. Caribou, reindeer, and musk ox calves run with the herd as soon as they have dried off. Pregnant females may remove themselves temporarily from the herd, like the plains bison that seeks a secluded area such as a coulee, behavior dating back to the natural days of wolf and plains grizzly bear predation; but in a few hours the calf is up and running and rejoins the herd with its mother, and nibbles grass within a day or two of its birth. On the high Andean grasslands, or altiplano, young guanaco and vicuna run fast and keep up with the herd within hours of their birth. Newborn wild Bactrian camels can walk two hours after they are born and follow the herd the next day. All wild sheep lambs and goat kids are highly precocial and follow their mothers along the most precipitous crags soon after birth. On the East African savannah, calves of the migratory species, such as the gnu, hartebeest, topi, and bontebok, have very precocial calves, allowing the herd to stay on the move. When they keep together, the calves are protected by the adults around them, but hyenas try to separate them and pick off the unprotected young. The foals of all the wild equids (zebras, Przewalski horses, and asses) are up and running within an hour of their birth. The young hippopotamus may be born under water and swims immediately, and it may also suckle underwater. Unlike the altricial swine (*Suidae*), young peccaries (*Tayassuidae*) are well haired, follow their mothers soon after birth, and return to the herd within a day or two.

Having a baby "at heel" increases the mothers' vulnerability to predation, but the highest risk from predators occurs at birth, when she is immobile and occupied. Herd animals highly vulnerable to predation reduce the risk through synchronizing their parturition. All the gnu or wildebeest on Tanzania's Serengeti Plains produce their calves within a few days of each other, and predation is very high until the lions and hyenas are satiated. If breeding occurred throughout the year, calf predation would be higher as it could occur more frequently. However, life in a large herd, with lots of calves being born in close proximity, creates another problem,

Collared Peccaries *Unlike the naked and virtually helpless piglets of the true swine, baby peccaries are born with a coat of hair, can run within a few hours of their birth, and soon return to the herd with their mother. They frequently reach her teats from the rear rather than from the side like most mammals.*
Photo: Kirill Troitsky, Shutterstock.com

that of recognition. Parents of the precocial runners have a strong infant-mother bonding mechanism that begins early in their lives, because the mobile nature of the runners obviously cannot succeed without an attachment between the mother and her baby. The maternal attachment happens quickly through smell, by licking and nursing her baby; but filial recognition may take a few days, and calves are rebuffed when they attempt to suckle from the wrong mother. Vision and sound are also involved in mutual recognition; eland recognize each other by their particular grunts and clicks. On cattle drives in the early American West, the newborn calves that could not keep up with the herd were carried in a calf wagon; and to prevent the mingling of their odors and confusing their mothers, they were each wrapped in their own sack during the day so they could retain their own body odor. In contrast, feral water buffalo cows in northern Australia adopt orphaned calves, and allow other calves with mothers to suckle.

Elephants are not ruminants but the sole members of their own family *Proboscidea,* survivors of giant herbivores that lived long ago, and they are often called subungulates. The condition of the young at birth is reminiscent of the ungulates, their babies being highly precocial and perfect miniatures of their parents. They can stand shortly after birth, and the herd waits for two days before they continue their search for food and water. All the herd members protect the babies from lions and hyenas in Africa and tigers in Asia. The rock hyraxes and tree hyraxes, said to be the elephant's closest relatives as they shared a common ancestor long ago, are also very precocial, able to run, jump, and climb trees soon after birth.

Most nonungulate runners are rodents, and most of those are South American. The 17 species in the order *Caviidae* have precocial babies, fully furred and mobile at birth. Their eyes are open and some eat solids within a few days. The guinea pig, long domesticated for food and more recently as a pet, is the most familiar of these. Other precocial rodents are the capybara—the world's largest rodent—the chinchilla, pacarana, spotted paca, agouties, hutias, and the coypu or nutria, whose nipples are situated so high on her sides the young can suckle in the water. To produce such well-developed young, these mammals have long gestation periods. The spotted paca's is 120 days, the chinchilla's 115 days and the pacarana's about 250 days, compared to the house mouse's 20 days and the brown rat's 21 days. Despite this long development period, however, they have only one or two young, sometimes three or four in the chinchilla, compared to the large litters of naked and helpless pups of most rats and mice.

■ HIDERS

Although it may seem more opportune for the young of the open range species to run and the forest species to hide, this is not always the case, and several grassland ungulates hide their babies for some time after birth. If the baby cannot initially keep pace with its mother or the herd, then it must hide. On the remnants of the North American prairie, the pronghorn (*Antilocapra americana*) leaves its calf for several days, and the blackbuck (*Antilopa cervicapra*), now a rare antelope of India's central plains and more common in Texas to where it has been introduced, hides its calf in tall grass for its first week. Their mothers return periodically just to nurse them. But it is the forest and bush species rather than grassland ungulates that are more likely to hide their young. In Africa, several ungulates of the bush hide their babies, and there are many variations to this behavior. The nyala of southeastern Africa gives birth in a thicket, and her calf remains hidden there for at least two weeks before venturing out. The klipspringer frequents bush-covered rocky areas, especially the stony outcrops called "kopjes," and actually hides her baby for three months. The male stands guard when she returns twice daily to feed it. The giraffe, the largest ungulate of the acacia-studded grasslands and mopane scrub, hides its calf for several weeks. The female roan antelope (*Hippotragus equinus*) separates from the herd to give birth and then stays with her calf for five days. She then returns to the herd but leaves the calf in a secluded place. It hides there for about four weeks before returning with its mother to the herd, where it joins a crèche of other mothers and babies. But they are very vulnerable to predators, and in Kruger National Park there is a mortality rate of 80 percent of roan calves in their first two months.

Leaving their calves in a crèche or nursery group is common behavior for several ungulates, including the greater kudu (*Tragelaphus strepsiceros*), eland (*Taurotragus oryx*), impala (*Aepyceros melampus*), and waterbuck (*Kobus ellipsiprymnus*). The babies spend a lot of time grooming and licking each other and developing strong bonds. Females accompany their unweaned calves in the groups, or they leave them and return periodically to nurse them. Eland mothers stay with their calves in the nursery group only until they are weaned at three months old, but the calves stay on until they are mature.

In the dense rain forest of equatorial Africa, the newborn okapi (*Okapia johnstoni*) follows its mother for the first day of its life, then finds a hiding place and remains there for the next eight weeks, being visited by the mother only for nursing. The other large forest antelope, the bongo (*Tragelaphus euryceros*), hides its calf for only one week, and then takes it to the nursery herd. Like the thousands of gnu on the Serengeti Plains that overwhelm predators with the abundance of prey when they all give birth close together, forest ungulates also have a similar survival strategy. North America's white-tailed deer (*Odocoileus virginianus*) have their fawns within a short time frame; and, like several other deer, including the moose, mule deer, roe deer, and the South American brockets, they hide their fawns initially.

Several rodents also hide their newborn young. It is unavoidable in the beaver (*Castor candensis*), giving birth in its winter lodge, safe from the cold. A monogamous species that lives in family groups comprising the mated pair and their young from their previous litter, the beaver's kits are born fully furred, with their eyes open, and can swim soon after birth. The father also helps with their care, grooming them and bringing food, but they remain in the lodge for about one month. Another North American hider is the porcupine (*Erethizon dorsatum*), whose baby is well developed at birth, with open eyes and long, black hair over short, soft quills. The

White-tailed Deer   *A typical "hider," a white-tailed deer fawn waits for its mother to return to feed it. The mother ate the afterbirth and then took the fawn to a safe place, where it will stay for up to two weeks. It lies very still, even in a very uncomfortable place, and lacks scent, so it is difficult for coyotes, cougars, wolves and bobcats to find.*

Photo: Bruce MacQueen, Shutterstock.com

baby hides on the ground initially, while its arboreal mother stays in the trees. She returns at night to feed it, and when about one week old it follows her into the branches, where it begins to eat solids and may be weaned at the age of two weeks.

Hares and jackrabbits are solitary animals and, unlike rabbits, do not dig burrows. They sleep in a grass-lined hollow or "form" in the grass or under a bush, and are well camouflaged when they lie flat and lower their ears. Their two or three well-developed babies or leverets are born in the form, and are fully furred and have their eyes open; but although they are mobile, they stay in the nest for about three weeks—when they are independent. They lie still and quiet until their mother calls when she returns to nurse them, and they answer to direct her to the nest.

■ MILK

The first milk produced by a new mother is called colostrum, and in the ungulates it is produced for only a few days after birth. It is rich in vitamins, especially A and D, and in minerals; and it contains the immunoglobulins (antibodies) that are absent in the newborn but are essential to protect it against infections of the digestive tract and to assist in developing its immune system. Colostrum has a low lactose content, as this can cause diarrhea or scouring. It is important that the baby ungulate gets colostrum for its first 24 to 36 hours, after which the stomach enzymes degrade its antibodies.

Milk varies in composition, and of course in quantity, depending on the size of the species. But it also varies according to the number of young the mother is feeding. For instance, it is known that some animals do not just produce a certain amount of milk and expect it to satisfy their young, however many are suckling, for studies have shown that ungulates with twins produce much more milk than others with just one calf to feed. Milk composition varies considerably, even in species with similar life styles. For example, red deer milk is rich in fat (11.5 percent) and protein (7.5 percent), whereas the milk of the wapiti and the white-tailed deer is more dilute, having less fat and protein. It also varies between domesticated animals and their wild relatives. Horse's milk has just 1.6 percent fat and 2.7 percent protein, while the milk of Burchell's zebra has 4.6 percent fat and 2 percent protein.

There is obviously a relationship between hiding and being nursed infrequently and therefore needing rich milk, compared to "running at heel" and being able to suckle more often. Mountain goats, which stay close to their mother for protection, have been observed sipping up to 36 times in one day. Their milk is quite dilute, containing 3.3 percent fat and a similar percentage of protein. In contrast, the milk of the giraffe, which hides its calf, contains 12.5 percent fat and almost 6 percent protein. Like the milk of most if not all mammals, the composition of ungulate milk changes during the lactation period, although the changes are nowhere near as dramatic as in the milk of the marsupials. When the elk calf is three months old, its mother's milk increases in fat from 7.5 percent to 10 percent, and in protein from 6.2 percent to 8.5 percent.

Milk composition may also be affected by other factors, that have nothing to do with the offspring's development, hiding or running. The composition of camel milk, for example, depends on the animals' water intake, as this affects the fat content. On a long desert journey, when they have not drank for several days, their milk is very rich, but after drinking their fill (perhaps 25 gallons [100 L] in 10 minutes), camel milk has a very high water content and just 1 percent fat.

Vertebrates cannot digest plant fiber—the structural components of plants such as lignin and cellulose—as they do not possess the correct enzymes. The ruminants have therefore evolved a symbiotic relationship with their intestinal microflora (bacteria and protozoa) that perform this task in the animals' rumen, serving their own purpose but also assisting the host by producing as a by-product the fatty acids it can digest. The young ruminant digests its milk in its acidic stomach, or abomasum. It would be dangerous for the milk to enter the fiber-fermentation chambers of the rumen and reticulum, so it bypasses these through a channel called the reticular groove that directs it into the abomasum. Before a young ruminant can begin to ruminate, however, its rumen must be inoculated with microflora, and this occurs while the infant is suckling, and also from its mother's licking, as her saliva contains the bacteria. It may also be acquired from the environment, as deer fawns removed from their mother immediately after birth developed normal rumen activity.

■ SOME OF THE SPECIES

Giraffe (*Giraffa camelopardalis*)

The most recognizable ungulate, and the tallest land mammal, mature bulls reach a height of 19 feet (6 m) and weigh 2,150 pounds (975 kg), but like most mammals their necks contain only seven cervical vertebrae that are obviously very elongated. Females are smaller, reaching a height of 15 feet (4.5 m). Giraffes have a tufted tail, and large feet, with the third and fourth digits bearing hooves. Both sexes have a pair of short, tufted horns, arising above and behind the eyes, with a bony core covered with skin and hair that develops unattached and then fuses to the skull as the animal matures. There is only one species of giraffe, with several subspecies that differ mainly in their patterning and have a geographical distribution that originally covered most of Africa in suitable habitat, but is now considerably reduced. They live in dry woodland, mopane scrub, and wooded savannah, where they browse the leaves of the acacia trees.

Giraffe calves are hiders, and parental care is totally the mother's responsibility. Initially the calves hide in thick bush while their mothers feed; then after several weeks they join the herd and are left in a crèche with other babies while their mothers go off to browse. Occasionally an adult supervises them, but they are usually left on their own and are therefore very vulnerable to the larger predators—lions, leopards, hyenas, and African hunting dogs. The calves stay in the crèche for several months and then accompany the adults, by which time they are quick enough to escape most predators. They prefer to stay in more open country, where lions cannot creep up on them unseen, but as the trees are sparse they must move around a lot to find sufficient food. Even so, predation of young giraffes is very heavy, and in some regions only about 25 percent of the calves reach maturity.

Blackbuck (*Antilope cervicapra*)

Blackbuck are the formerly common gazelle-like antelope of southern Asia, but they were exterminated in many parts of their range in the twentieth century, through hunting and habitat destruction. However, they were introduced into Texas in 1932 as game animals and flourished there, and some have since been returned from there to their native India. They are polygynous animals; the females live in small herds and are mated when they cross over the territories of the dominant males. Adult males are blackish brown above, and the females yellowish brown, and in both sexes the color extends down the outsides of their legs. Their bellies and insides of their legs are white, and they have white eye circles. Young bucks resemble the females in color, begin to darken when they are 15 months old, and attain their adult color at four years of age. Adult males weigh up to 80 pounds (37 kg) and reach a height of 30 inches (76 cm) at the shoulder. Their long, V-shaped spiral horns may be 31 inches (79 cm) long. Blackbuck are almost exclusively grazers of short grasses on the open plains and dry thorn forest of India and Pakistan. Like all grassland antelope they are very fast, and can be outrun only by the cheetah, which is now extinct in the wild in Asia. Blackbuck calves are hiders, and for their first week they lie hidden in tall grass and undergrowth. This is a very dangerous time for them, with tigers, leopards, and wild dogs or dholes searching for them, and a mother visits her calf only to nurse it. At the start of their second week of life they accompany their mothers and rejoin the female herd, where they remain until they are independent.

White-tailed Deer (*Odocoileus virginianus*)

The most common New World deer, the white-tailed deer has been hunted extensively by the native people of North America since their arrival from Asia, and is a prime target for sport hunters. It is widespread in a variety of habitat, from southern Canada through the United States and Central America to northern Brazil, excluding the arid lands of the American southwest as well as the Pacific Northwest, where it is replaced by a subspecies of the mule deer. Within its large natural range, the many subspecies show considerable variation in size. Northern deer may weigh 330 pounds (150 kg), and stand 39 inches (1 m) at the shoulder; tropical rain forest animals may be 32 inches (80 cm) tall and weigh 120 pounds (55 kg), whereas the smallest race of all lives in Florida's Everglades and the Keys, and is only 26 inches (65 cm) high and weighs 80 pounds (36 kg). The white-tail is reddish brown in summer, brownish gray in winter, and has a coat of brittle, tubular hairs. It resembles the mule deer, the most obvious differences being the color of the tail, which is brown above and white below, and its smaller ears—which are just half the length of the head. The white-tail's antlers have a single main beam and several smaller branches. Its fawns are very vulnerable to cougar, lynx, bobcat, wolf, and coyote, but it has evolved several survival strategies to counter them. Most females give birth within a short time frame in what has been termed "predator swamping," when there are more fawns about than the predators can consume. In addition, the fawns are hiders; to make them difficult to find, their mothers eat the afterbirth, their spotted coats camouflage them in the sun-dappled woodland, they lie very still with their neck outstretched, and they lack scent. The mother eats their feces, which they do not pass until she arrives to feed them every six hours. In more populated and visited areas, these survival strategies now often backfire, when people "rescue" fawns they believe are abandoned.

African Elephant (*Loxodonta africana*)

The African elephant differs from its Asian relative in a number of ways. It has larger ears and a flat and slightly backward-sloping forehead. It has high shoulders and rump with a dip between them, unlike the domed back of the Asian elephant. Its trunk ends in two finger-like extensions, and it has 21 pairs of ribs and 26 caudal vertebrae. Bulls have very long and heavy tusks, whereas those of the females are thinner and shorter. Two subspecies are usually recognized—the savannah or bush elephant (*L. a . africana*), and the forest elephant (*L. a. cyclotis*). They apparently do not interbreed even though their ranges overlap where the forest meets the savannah. The forest elephant is the smaller of the two, and has more rounded ears, smaller and more slender tusks, and a darker skin. Mature bulls stand 9 feet (2.8 m) high at the shoulder and may weigh up to 11,000 pounds (5,000 kg); females are smaller. Its habitat is the rain forest zone of west and central Africa. The savannah elephant is the world's largest land mammal, weighing up to 15,400 pounds (7,000 kg), and with a shoulder height of 13 feet (4 m). Its head and body length —including the trunk—may reach 23 feet (7 m). Its skin color has a more brownish tone than the forest elephant, and it has exceptionally large ears, measuring up to 6 feet (1.8 m) from top to bottom. The savannah elephant now occurs only in protected areas of Africa beyond the tropical forests, preferring wooded savannah, acacia, and mopane scrub. It has a dramatic effect on the African bushveldt, its great appetite and strength profoundly affecting vegetation and in turn influencing the lives of many other species. Tearing off large branches, uprooting trees, and killing others by bark-stripping, elephants have made wooded savannah more vulnerable to fire, and to more intense fire, further destroying trees and creating open grassland.

Elephants have the most extended period of parental care of all the land mammals, excluding the great apes, for calves depend upon their mothers for up to five years. They can stand within an hour of birth, but the herd waits for two days until the baby can keep up as it moves and forages. Calves nurse solely for their first two years, then are weaned slowly, but they may continue to suckle until the birth of the next baby. Elephants live in permanent family groups, matriarchal societies in which all the members are very affectionate to the babies and protect them; whereas adult males are solitary. The strong bond between family members provides the young with opportunities to interact with their older siblings and adults, who also care for them—behavior known as allomothering—helping to maintain the society.

Burchell's or Plains Zebra (*Equus burchelli*)

Excluding the narrow-striped Grevy's zebra and the mountain zebras, all zebras are considered races of Burchell's zebra, also known as the plains zebra—the most common wild horse in Africa. They have vertical stripes from their head to

African Elephants *The herd is very protective of the latest baby, and waits for two days after its birth before continuing their perennial search for food and water. There is a strong bond between the family members in this matriarchal society, and older siblings and other females all assist the mother in caring for her calf, which will be dependent on her for milk for at least five years.*
Photo: J. Norman Reid, Shutterstock.com

mid-body, where they begin to turn towards the tail and become horizontal on the rump. Although broad and well-spaced rump stripes are characteristic of the species, there is considerable variation in the several subspecies plus geographic differences and individual aberrations. These include some animals with striping down to the hooves, others in which leg stripes are almost absent; others with faint or "shadow" stripes between all their stripes, or just between those on the rump, or no shadow stripes at all. Their stripes go around the belly and meet at the mid-ventral line, unlike Grevy's zebra, whose stripes do not meet under the belly. Reverse-patterning occasionally occurs, in which a zebra has a black coat with white stripes. Burchell's zebra is a compact horse, standing 55 inches (1.4 m) at the shoulder and weighing up to 825 pounds (375 kg). It is an animal of the savannah, scrub, and open woodland, where it is mainly a grazer but browses when grass is unavailable; and it may be sedentary or migratory depending upon the availability of food and water. The plains zebra is polygynous; a stallion controls a herd of several mares and their foals, and drives young males away as they approach sexual maturity. The foals are very precocial; they can walk within half an hour of their birth, and run within an hour. They must keep up with the herd as it moves to find grazing and water, and need its protection from predators.

Musk Ox (*Ovibos moschatus*)

Musk oxen do not have musk glands and they are not oxen. They have characteristics of both the cattle and sheep but are believed to be closer relatives of the takin, a large herbivore of central eastern Asia. They are very hardy animals native to the tundra of Alaska, Canada, and Greenland, where they are protected by their coarse outer coat of dark-brown guard hairs and an insulating inner lining of dense and soft light-brown wool called qiviut, and are impervious to the coldest weather. Musk ox are massive in form, and their bulk is enhanced by a shoulder hump and their long, thick coat that reaches to the ground. Both sexes have pointed curved horns, but the males' are much larger and are joined by a heavy boss. Adult bulls can weigh up to 1,325 pounds (600 kg) and stand about 52 inches (1.3 m) at the shoulder. They appear ponderous, but this is misleading, for they are very fleet-footed and move swiftly and with agility when necessary. Accustomed to bone-jarring head-on clashes with challengers, captive bulls have butted reinforced concrete walls without harm. Musk oxen seek moist habitat in summer, usually in river valleys or along lake shores where they graze on grasses and sedges, and swim well when necessary. For the winter they migrate to more exposed areas where the wind blows the snow off their foods—mainly ground-hugging browse such as willow and rhododendron. Freezing rain is hazardous, as it accumulates in their coats, chills them, and affects their mobility and effective defense against wolves, their major predators. Their calves are born with a thick coat of qiviut and run fast soon after birth. They shelter under their mother's long coat during severe weather. When wolves threaten the herd, it forms a protective circle around the calves, that the predators find very difficult to penetrate.

Bighorn Sheep (*Ovis canadensis*)

One of the largest of the true sheep, bighorns are heavily built animals with a brown coat, white rump, and belly. They have huge curling horns that are very thick at the base and curl around and upwards before turning outwards at their tips; and in adult males they can reach a length of 43 inches (110 cm) along the spiral. There are several subspecies, but there is disagreement over the number; some taxonomic authorities list only three, others several more. The subspecies vary considerably in size, but the largest is the Rocky Mountain bighorn (*O. c. canadensis*), in which adult males weigh up to 308 pounds (140 kg) and stand over 39 inches (1 m) at the shoulder. Bighorn sheep are animals of the alpine meadows, grassy slopes, and cliffsides of the mountain ranges of western North America, and are mainly grazers. They are social animals, and several ewes and lambs live together in summer and are joined by rams in late fall for the rutting season. The rams also live in flocks in which dominance depends on age, strength, and horn size, and their ritual contests include head-on charges. Loss of habitat, hunting, and disease introduced by domestic sheep has seriously reduced their numbers. The easternmost subspecies, from the Black Hills of South Dakota (*O. c. audubon*), was exterminated in the nineteenth century, but bighorns have been reintroduced there. Bighorn ewes find a protected place to give birth, and their single lambs are very precocial. Agile and sure-footed, they follow their mothers across precipitous cliffs soon after birth. The lambs join a crèche when they are four weeks old, and seek out their mothers when they need to suckle. They generally stay with their mothers until the birth of the next lamb the following spring.

White-lipped Peccary (*Tayassu pecari*)

A slightly larger animal than the more familiar collared peccary, the white-lipped peccary has a blackish-brown coat and white on the sides of the jaws. It reaches a maximum body weight of 88 pounds (40 kg) and stands about 22 inches (55 cm) at the shoulder. It has a similar New World range, although beginning further south, from southern Mexico and not Texas, and extending down to northern Argentina. It lives in desert scrub and woodland, but is primarily an animal of the tropical rain forest and is seldom found far from water. It has more migratory habits than the collared peccary, and ranges over a wide area. The white-lipped peccary is a very gregarious animal, living in herds of over 100 individuals containing males and females of all ages. Very aggressive in defense of their young, collectively they are able to chase away jaguars and pumas. They are mainly vegetarians, eating leaves and shoots and digging in the loose leaf litter of the forest floor for roots, fallen fruit, fungi, seeds, nuts, and invertebrates. White-lipped peccaries generally have twins or triplets, and the pregnant female finds a secluded place in dense undergrowth to give birth. Her young are well haired at birth and can run within a few hours. She takes them back to the herd the following day, where all its members are very protective.

Patagonian Cavy (*Dolichotis mara*)

A large slender-legged, harelike rodent, the Patagonian cavy, or mara, walks, hops, gallops up to 27 miles per hour (45 kph), and bounces on all fours in a stiff motion called "stotting," similar to the pronking of some antelope. They rest in a catlike sitting position, with their front limbs turned inwards beneath the chest. When adult they are about 29 inches (75 cm) long and weigh up to 33 pounds (15 kg). Their backs are dark gray, they have buffy-brown bellies separated by a broad band of pale gray, and there is a very distinctive white band around the rump. Maras are natives of central and southern Argentina, where they live in the semiarid regions and dry grasslands, but they are declining and locally rare due to hunting and competition from the introduced European hare. They have been introduced into Brittany, where they are established even though they still breed mainly in the cold and wet winter, which coincides with Argentina's austral summer. Mara are monogamous, generally mating for life, and live in burrows of their own digging or enlarged armadillo holes. Their reproduction involves an unusual combination of precociality and hiding. Females are sexually mature at six months, and three or four young are born after a gestation period that has been variously recorded from 76 to 93 days. The babies are highly precocial and mobile at birth, are fully furred, and have their eyes open. Although they are usually born outside the parent's burrow, they enter a communal crèche burrow containing other young, and remain there for several weeks. They begin to come out briefly to nibble grass when just two days old, and their mothers visit the den several times daily to nurse them.

Notes

1. Hair is used here in a general sense, to include also fur (a coat of fine, silky hair) and wool, which is distinguished from hair by its covering of minute overlapping scales that soften the fiber.

11   Beach Babies

There are two groups of mammals whose young are precocial or well developed at birth. The preceding chapter included those who follow their mother soon after birth, or are hidden until they can. This chapter is concerned with the seals, sea lions, and their relatives, whose precocial pups are born on beaches, rocky islets, or polar pack ice. They are known collectively as pinnipeds, after their zoological order *Pinnipedia,* a name derived from the Latin for "fin-foot." They have been called aquatic mammals, but they are not totally dependent on water, although they do spend a lot of time there. With few exceptions, they mate on land or ice, and about one year later they bear and raise their young there, a completely different lifestyle from the totally aquatic whales and dolphins. A more appropriate description of their lifestyle is "semiaquatic."

The pinnipeds evolved from land carnivores in the Northern Hemisphere about 30 million years ago, and DNA analysis recently determined that they had a common ancestor. After evolving in northern seas, they began moving into southern waters and eventually occupied the coasts of the southern land masses and Antarctica about two million years ago, developing into new species there such as the crabeater seal, leopard seal, and the Weddell seal. On their way south, a few stayed in the tropics and became the monk seals of the Hawaiian Islands and of the Caribbean and Mediterranean seas. The pinnipeds are primarily marine animals, occurring in all the oceans of the world except the Indian Ocean; but there are four freshwater seals. The Baikal seal (*Phoca sibirica*) occurs only in freshwater Lake Baikal in southeastern Siberia; a subspecies of the harbor seal (*Phoca vitulina mellonae*) is confined to Seal Lake in Quebec; and there are two freshwater subspecies of the ringed seal (*Phoca hispida*), one in Lake Ladoga in northwestern Russia and another in Lake Saimaa in neighboring Finland. Pinnipeds range in size from the Baikal seal and Caspian seal (*Phoca caspica*), in which the average adult male weighs about 154 pounds (70 kg), to the enormous southern elephant seal (*Mirounga leonina*) that can weigh 4,800 pounds (2,200 kg). They all breathe air, but can deflate their lungs, dive deep, and stay submerged for long periods. The northern elephant seal (*Mirounga angustirostris*) holds the world record for deep-diving pinnipeds, going down 5,015 feet (1,540 m), and it can stay submerged for up to two hours.

The pinnipeds are divided into three distinct groups or families—the true or earless seals, the eared seals, and the walruses, with a total of 34 species. There are 19 species of true seals in the family *Phocidae*. In these animals. the rear flippers are used for propulsion, but they point backwards and cannot be rotated forward like the sea lions, so they move on land with an undulating motion called "gallumphing." They lack external ears and have thin fur that does not trap air, and are therefore more dependent on blubber for maintaining their body temperature. The family includes the tropical monk seals, northern species like the gray seal and harbor seal, and the Baikal seal. It also includes those species known as "ice" seals, which live in both polar regions and must haul out onto ice to give birth. They include the crabeater, Weddell, and Ross seals of the southern oceans and Antarctica, and the hooded, ringed, harp, and ribbon seals of the northern waters and the Arctic. The largest pinnipeds—the two species of elephant seals, one northern and one southern—are also members of the family *Phocidae*.

The second major group of pinnipeds are the 14 species of eared seals—the fur seals and the sea lions—that are included in the family *Otariidae*. They are all beach-breeders, and males are considerably larger than females. They have small external ears, and they can rotate their hind feet forward and move fast on land; but unlike the true seals, they use their large front flippers to power them in the water. They have denser fur that traps air and helps to insulate them. They are divided into two groups, the fur seals and the sea lions. There are nine species of fur seals—one in northern seas, the others all in southern waters—and they have more pointed snouts than the sea lions and thicker underfur. There are five species of sea lions: the northern or Steller's sea lion, the Californian sea lion, the South American sea lion, the Australian sea lion, and the New Zealand or Hooker's sea lion. They all have blunt snouts and a coat of sparse underfur covered by short, coarse guard hairs.

The walrus is the only member of the family *Odobenidae*. It can rotate its hindlimbs forward and beneath it, like the eared seals, and can therefore use them for land movement. It also uses both its front and back flippers for propulsion in the water. The walrus differs from all other pinnipeds in its possession of tusks, elongated upper canines that reach 36 inches (90 cm) in old bulls, and slightly less in females. It lacks external ears and has tough, wrinkled, almost hairless skin 1½ inches (4 cm) thick, and an underlay of blubber—the layer of fat beneath the skin of marine mammals. The walrus breeds on beaches and pack ice.

All the pinnipeds are carnivorous, their diet comprising fish, squid, octopus, shellfish, and krill. The fur seals may eat birds, and the leopard seal preys heavily on penguins and also eats young pinnipeds. In turn, they are preyed upon by polar bears, killer whales, and great white sharks, plus humans, who have hunted them for their hides, meat, and blubber for several centuries and continue to harvest some species commercially—a highly criticized and controversial practice. In addition, they have had to contend with the reduction in fish stocks, entanglement in fish nets, and the periodic El Niños that have warmed the eastern Pacific and have seriously affected the populations of the South American sea lion (*Otaria byronia*), the California sea lion (*Zalophus californianus*) and the northern elephant seal (*Mirounga angustirostris*).

The mating systems of the pinnipeds differ between the two major groups. The fur seals and sea lions are polygynous, males mating with numerous females. They are also highly territorial, the bulls arriving at the breeding beaches first to establish their territories, and fighting viciously to assert dominance over their rivals. The females, pregnant from the previous year's mating, then arrive, give birth and mate again a few days later. There is considerable difference in size between the sexes, and adult bulls may be four times the size of the females. A bull may control and mate up to 40 females during the short breeding season and usually fasts during this period.

The mating systems of the seals are more variable. Several, such as the ringed seal and hooded seal, are monogamous; whereas the elephant seal and gray seals breed in large herds. Size difference between the sexes in the seals also varies. The males and females of monogamous species are of similar size, because large male size is unnecessary when they do not have to compete for a territory. In contrast, the gregarious elephant seal bulls reach a tremendous size due to the need to compete, and fight viciously on the breeding beaches to rank themselves in a hierarchy. The walrus is also gregarious and lives in huge herds outside the breeding season, on the coastal shores and pack ice over the continental shelf in the Arctic Ocean. It has an unusual polygynous breeding system, similar to the lek system practiced by several birds. Females form large groups on the ice, and the males swim past and attempt to entice them to enter the water and be mated.

■ PARENTAL CARE

All pinnipeds produce well-developed and totally precocial young. They are furred, their eyes are open, and they can thermoregulate and maintain their body temperature. Birth normally occurs on a traditional calving beach, on the pack ice, and occasionally in the water. Parental care is totally the mother's responsibility and varies from just a few days in some polar seals, to over a year for the fur seals and sea lions, and two years for the mother walrus. Birth in the pinnipeds is generally quick, on account of their "sausage" shape, and delivery usually takes less than five minutes. Unlike most mammals, they do not eat their placenta. The body composition of the fetus varies according to the species, but at birth seal pups contain up to eight times more fat than the average mammalian baby, an evolutionary requirement to combat their sudden exposure to low environmental temperatures.

Walrus pups are usually born in the water, and California sea lion and harbor seal babies may be, but births generally occur on land or on the ice. Although the babies are mobile, they normally do not move far from the spot where they were born for their first few days. They have a thick coat, white in the harp seal, ribbon seal, and ringed seal pups; whereas hooded seal pups at birth are blue-gray on their backs and silver-gray or yellowish on their bellies. Young elephant seals have a brown coat and the newborn harbor seal is bluish gray with a whitish belly. Steller sea lion and California sea lion pups are dark chocolate brown in color, and when wet appear almost black. In most species their birth coat or "baby fur" is soon shed, however. Harp seal pups shed their white coats when two weeks old, and the ribbon seal pups at five weeks old. Hooded seal and bearded seal pups actually shed their white baby fur, or lanugo, in the uterus, and are born with darker coats.

Although newborn pinnipeds can swim, most do not enter the sea for several weeks; some have inadequate blubber until then and therefore have little buoyancy or insulation. Pups that are born on the ice pack—such as the harp, ribbon, and ringed seals—do not enter the water until they have molted their baby fur. Bearded seal pups enter the water soon after they are born. Some may nurse in the water; walrus pups do so from birth, and sea lion and fur seal pups generally later in their nursing period.

Seal milk is very rich, with a fat content of over 50 percent in some species. Protein is also well represented in their milk, but lactose is totally lacking (in sea lion milk) or is very low (up to 2.6 percent in seal milk), as the abundance of fat makes any other form of energy superfluous. The milk composition of the fur seals and sea lions remains almost constant during their long lactation period, yet despite their

South American Sea Lion Rookery   *A calving beach, or "rookery," of the South American sea lion on an island off the coast of Peru. Shiny black pups frolic in the shallows, awaiting the return of their parents from fishing in the once fish-rich waters of the cold Humboldt Current, now depleted by overfishing. It has also been seriously affected by the periodic El Niños.*
Photo: Clive Roots

much shorter nursing cycle, seals' milk increases in fat during the lactation period. The composition of their milk reflects the different nursing strategies. The fur seals and sea lions nurse their pups for several months and have more dilute milk. The seals have a relatively short lactation and nursing period, and have fat-rich milk.

The hooded seal produces the richest milk of all mammals, with a fat content of 63 percent, 18 times richer than cow's milk, but only 6 percent protein. Elephant seal milk contains 55 percent fat, the common or harbor seal's milk has 45 percent fat, and gray seal milk has 53 percent fat. The crabeater seal's milk has 51 percent fat, and its pup may drink as much as 158 pints (90 liters) during the lactation period. Bearded seal pups drink about 14 pints (8 liters) of milk (containing 50 percent fat) each day and double their birth weight of 77 pounds (35 kg) during their nursing period of 21 days. Harp seal milk contains 50 percent fat, a diet on which the pup gains 70 ounces (2 kg) per day.

The pinniped's lactation periods vary. It is very short in the seals, with the hooded seal abandoning her pup after nursing it for just four days, the shortest of any mammal. But during those days it increases its weight from 46 pounds (21 kg) to 94 pounds (43 kg), most of this increase being fat. Harp seal pups suckle for only 12 days, crabeater seals nurse for 14 days, the gray seal weans its pup when it is 18 days old, and the harbor seal leaves her pup when it is between four and six weeks old. Surprisingly, the Baikal seal feeds her pup for eight weeks. On the beaches of California and Mexico, the northern elephant seal feeds its pup for three weeks before returning to the sea for its long migration north to the Gulf of Alaska.

The differences in milk and nursing periods are summed up as follows. Female seals give birth on land or on the ice, and then fast while producing very rich milk for their pups, which are weaned between four days and eight weeks according to the species. Sea lions and fur seals also give birth on land and initially stay with their pups, fasting and nursing, but then alternate time at sea feeding with time ashore nursing. Their milk is not as rich, but they have a long lactation period, ranging from four months to two years. Walruses give birth on ice floes, and sometimes in the water, and then bring their pups ashore. They fast on land for a few days, while feeding the pups, and then return to the sea with their pups and nurse in the water. Their lactation period lasts up to two years, and their milk is relatively low in fat. It is therefore a fallacy that all pinnipeds have short nursing periods, but they do indeed all have rich milk, and the seals have very rich milk. Although water is the major component of the milk of other mammals, the solids (fats and proteins) in seal milk exceed the water content. It is likely that milk fat percentage is influenced by diet with variations due to the seasonality of prey abundance, with very oily fish resulting in an increase in the milk's fat content. Seal milk also tends to increase in fat content toward the end of the lactation period.

In most mammals, weaning means changing slowly from a diet of milk to one of solids, but in the seals weaning is abrupt. Mother seals do not bring food back to their young to wean them, nor do they take them into the sea where they can learn to fish by example while still being suckled on land. They simply abandon them, but they have prepared them well for the sudden end to this brief period of their lives. The pups have stored a large amount of fat during their short nursing period and this is metabolized during their post-weaning fast. They do not go to sea immediately, and survive off their stores until they do. Hooded seal pups wait about two weeks, but the northern elephant seal pups may lie on the beach for two months before taking to the water. Harp seal pups, left on the ice 12 days after birth, do not swim until they are one month old. Learning to fish for themselves, however, is not easy. Gray seal pups are known to catch easier prey such as crabs and shrimps until they become expert at fishing, but while they are learning to survive they may lose one-third of their body weight.

Female fur seals, sea lions, and walruses have a totally different nursing strategy. After giving birth and being mated again, they alternate short periods of suckling their young with much longer periods, usually of two or three days, at sea feeding; a pattern that continues for several months. The northern fur seal pup is suckled for four months; but in the other fur seals and the sea lions, nursing may continue for over a year, and as they accompany their mothers the growing pups follow their example and learn how to fish.

Several species of fur seals and sea lions practice alloparenting—when a mother provides care for a baby that is not hers—as they are known to adopt and feed orphaned or lost pups. This behavior occurs in harem species, where the young in the group all have the same father, and caring for the young of related females improves their chances of survival and thus provides genetic advantages. However, it also occurs in unrelated animals, when it seems to lack any advantages. Just the opposite, in fact, for it uses energy to ensure the survival of the offspring of rival males, and the reasons for this are unclear. While their mothers are away at sea, pups may band together in a crèche, supervised by a nonbreeding female who protects them and herds them back to the shallows from deep water. Bulls have also been seen guarding the females and pups in their herd, warning them of sharks and even chasing sharks, the closest pinniped males come to providing care. In contrast to the unselfish alloparenting behavior of the females, in the harem-forming species pups may be harassed, injured, or even killed by unrelated and unattached males, generally subadults that enter a bull's territory during his absence. This happens mostly towards the end of the breeding season, when bulls are perhaps less watchful. Pups also risk their lives if they stray into another bull's territory.

■ SOME OF THE SPECIES

Northern Elephant Seal (*Mirounga angustirostris*)

This is the largest pinniped, in which adult males weigh up to 5,070 pounds (2,300 kg) and are 15 feet (4.5 m) long. The females are much smaller, weighing only 1,650 pounds (750 kg) and reaching a length of 12 feet (3.6 m). In addition to their huge size, they have a distinctive snout that droops over their muzzle. Their blubber, meat, and pelts were very hard to resist, and they were hunted almost to extinction in the nineteenth century, but have recovered with strict protection. They are migratory animals that make a return journey annually between the California and Mexico calving beaches and their feeding grounds in the north Pacific Ocean, especially around the Aleutian Islands. This long annual migration, similar to that made by the gray whale, totals 13,000 miles (21,000 km) and takes up most of the year. Their main foods are squid and octopus, plus fish, crabs, and even small sharks.

Elephant seals are gregarious and polygynous, a dominant male controlling a group of females and aggressively denying other males access to them. The males arrive first at the breeding beaches in December and fight for a high position in the hierarchy, allowing them access to more females. Both males and females fast while they are at the breeding beaches, and the females are mated soon after they have given birth. The gestation period is only nine months, but they practice delayed implantation for three months so that births will coincide with their return to the beaches almost one year later. Their pups weigh 66 pounds (30 kg) at birth and have silvery-black fur, but molt and replace this with a silvery coat when they are weaned at three weeks old.

As the females were mated while nursing, they are then free to return to the sea for the journey back to Alaska. Even though its milk is already very rich, the fat content of the elephant seal's milk actually increases during the last week of lactation, preparing the pups for their abandonment. Before going to sea, the pups live off their blubber for up to two months, and usually join a crèche of other babies for this period, but they may steal a drink from other females still nursing their pups.

Hooded Seal (*Cystophora cristata*)

Hooded seals live in the central and western North Atlantic, from the Gulf of St. Lawrence into Arctic waters north of Greenland. Like most pinnipeds, they are sexually dimorphic—there is a noticeable size difference between the sexes. Males are larger, and may weigh over 660 pounds (300 kg) and are 8.2 feet (2.5 m) long. They have an impressive "hood," a large nasal sac that covers the top of their muzzle from nostrils to forehead. It hangs over the muzzle when flaccid, but when they display and inflate it with air from the nostrils, it expands into a large red balloon. Females lack the hoods, weigh 440 pounds (200 kg) and are about 6 feet 6 inches (2 m) long. Their adult coat is silver gray with dark blotches. They are mainly fish-eaters, but also take shellfish and starfish, and in turn they are hunted by killer whales.

Hooded seals are solitary animals, except when they molt and breed. From all over their large range they congregate to breed at four major sites, and to molt at just one—off the western coast of Greenland. Pups, born on the pack ice, are called "blue-backs," having shed their fetal coat or lanugo before birth. They have a well-developed layer of blubber and have blue-gray backs with silvery-gray or yellowish bellies. Hooded seal milk is the richest milk of all mammals, with a fat content of 63 percent. Consequently, they have the shortest lactation period of all mammals, weaning their young after just four days of nursing, during which time the pups double their birth weight of 46 pounds (21 kg). The fat pups stay on the ice for at least two weeks after their mothers leave, living off their fat reserves, before they enter the sea for the first time. Hooded seal pups were once "harvested" annually for commerce, but commercial hunting of infant hooded seals is now banned in

Harp Seal *One of the ice seals, a "white coat" harp seal pup waits for its mother on the St. Lawrence River ice. Once the target of seal hunters, the harvesting of white-coated pups has been banned in Canada since 1987, but pups can be killed after they have molted their white coats. After nursing for only 12 days on milk containing 50 percent fat, they are abandoned by their mothers, but do not enter the sea until they are one month old.*
Photo: Photos.com

Canada, although adults may still be hunted. The United States maintains a complete ban on the hunting of all marine mammals.

Harp Seal (*Phoca groenlandica*)

Harp seals live in the North Atlantic and Arctic oceans, from the Gulf of St. Lawrence to northern Russia, and are associated with the pack ice. As it recedes, they head north to feed in Arctic and sub-Arctic waters during the summer, and then move south again in front of its advance in the autumn. There are three distinct populations, the largest being the northwest Atlantic herd with perhaps five million animals. They are a small species of seal, with males weighing up to 286 pounds (130 kg), and about 6 feet (1.8 m) long; the females are a little smaller. They are light gray with a large harp-shaped, dark-brown ring on their backs, and have black faces.

Harp seals gather in large herds to molt and then to breed, giving birth to a single pup on the pack ice in February or March. They are the most important commercial seal; their large numbers and appetites for fish that are coveted by humans, justifying the annual harvest. The pups have long, fluffy, white coats, but the commercial hunting of the infants, known as "white coats," was banned in Canada in 1987 after much international criticism. They may now only be killed after they have molted, which they do at the age of two weeks when their new coats are silver-gray with dark spots. They are hunted in Russia and Norway, but the largest commercial hunt occurs in Canada, where, in the 2005 killing season between late March and mid-May, 300,000 pups were harvested. Harp seal pups nurse for 12 days on milk containing 50 percent fat, and in that short time increase their weight from 22 pounds (10 kg) at birth to 77 pounds (35 kg). The mother fasts while she is lactating and may lose 6.6 pounds (3 kg) of body weight per day. During this period she is mated again, then leaves her pup and returns to the sea. The pups stay on the ice for one month before entering the sea themselves.

Harbor Seal (*Phoca vitulina*)

The common seal of the coastal waters and estuaries of the North Atlantic and North Pacific, the harbor seal is one of the smaller species, a frequent visitor to harbors and marinas, where it cruises slowly around with just its shiny, round head above the surface. Males are 6 feet (1.8 m) in length and weigh 250 pounds (113 kg); females are a little smaller. It has a fusiform shape—rounded in the middle and tapering at the ends—and is either light gray to silver with spots, or dark gray to brown with rings, but it looks very dark when it is wet. Its fore flippers have claws that are used for scratching and grooming, and it propels itself with its hind flippers, moving them side to side and using them as a rudder. Unlike the sea lions, it cannot rotate the hind flippers beneath its body, and it moves on land in an undulating caterpillar-like motion. It eats squid, mollusks, crustaceans, and a wide variety of fish. It does not chew, but swallows its food whole or bites off chunks, and crushes shellfish with its molars.

Harbor seals are solitary animals that only interact to mate, when the males each mate with several females. Their gestation averages 10 months; and the single pup, weighing about 24 pounds (11 kg), is born in spring or early summer, on land or on the ice or even in the water, as it can swim at birth. Harbor seal milk is very rich and contains 45 percent fat, 9 percent protein, and a trace of lactose. The pups nurse, on land or in the water, for a period of four to six weeks, when they are abruptly left by their mother. About 1,000 harbor seals of the subspecies *Phoca vitulina mellonae* live in landlocked, freshwater Seal Lake in Quebec. How they got there is a mystery.

Baikal Seal (*Phoca sibirica*)

Lake Baikal is an isolated freshwater lake in southeastern Siberia, yet it has a population of small seals whose closest relatives are the ringed seals living in the Arctic Ocean, 1,200 miles (1,930 km) to the north across the forests and tundra of Siberia. They are the only species of pinniped occurring solely in freshwater, and are believed to have been isolated in the lake for 500,000 years, but how they reached it is unknown. The lake is the world's largest body of freshwater, containing 25 percent of all the world's surface freshwater; with a depth of 5,370 feet (1,637 m), it is the deepest lake on earth. It is 394 miles (636 km) long by 43 miles

(70 km) wide. When mature, male Baikal seals are 55 inches (1.4 m) long and average 154 pounds (70 kg). They are often said to be the world's smallest seals, but they share this distinction with the similarly sized Caspian seal (*Phoca caspica*), another inland sea species, although its sea is slightly saline. They have a dark silvery-gray coat with paler, yellowish-tinged underparts. They are solitary animals that spend most of their time in the water, and in winter they have breathing holes and haul-out holes in the thick ice. Baikal seals have no predators in the water, but brown bears occasionally ambush a seal along the shoreline. They are mainly fish-eaters, especially of the salmon-like omul (*Coregonus autumnalis migratorius*) and the golomyanka (*Comephorus dybowksii*), both endemic lake species like the seal.

The current population of Baikal seals is about 85,000, but there have been heavy losses in recent years due to pollution and hunting, with about 10,000 killed annually. Organochlorines and other chemical pollutants in the food chain, especially from lakeshore pulp mills, are affecting their reproduction, and canine distemper virus killed thousands recently. The pups are born in late winter in dens hollowed out of the ice, where they are safe from attack by crows. They weigh 8.8 pounds (4 kg) at birth, and have a white and woolly coat, which they molt at seven weeks, and are then silvery gray on their backs with yellowish-gray bellies. They are weaned when eight weeks old.

Weddell Seal (*Leptonychotes weddelli*)

The Weddell Sea is part of the Southern Ocean bordering the Antarctic Peninsula, and much of it is covered with permanent shore-fast ice. Named for British sailor Captain James Weddell, who entered the sea in 1823, it is where Shackleton's ship the *Endurance* was crushed by sea ice in 1915. It is also the home of a very unusual seal, a large pale-gray animal with many small spots and blotches in which, in a reversal of typical pinniped sexual dimorphism, the females are larger than the males, weighing up to 1,100 pounds (500 kg), the males about 990 pounds (450 kg). They have a short, dense coat that protects them from the cold water, which is usually just above the freezing point of sea water at 29°F (-1.6°C). Their other unusual feature is the smallness of their heads for such large bodies.

Weddell seals live close to the Antarctic coast and do not migrate north in winter to escape the intense cold, but spend most of their time under the thick pack ice. They must therefore keep breathing holes open, which they do with their teeth, even when the ice is 6 feet (1.8 m) thick, and the damage to their teeth and gums through ice-chewing shortens their life span. They are great divers, going down 2,000 feet (610 m), and staying submerged for one hour; and may swim several miles from their breathing holes, using sonar to navigate, communicate, and possibly to locate fish. They are also assumed to have good vision for swimming under the ice in the poor light. They range farther south than the other Antarctic seals, reaching to within 800 miles (1,287 km) of the South Pole, and are therefore the world's southernmost mammals. They are mainly fish-eaters, especially favoring Antarctic cod, plus squid and crustaceans.

Weddell seal pups are born in colonies on the pack ice, weighing 65 pounds (29 kg) at birth and having their permanent teeth. They have a gray coat, but this is replaced at three weeks with a darker one. Their mother's milk contains 60 percent fat, and the pup is weaned at six weeks, by which time it has gained 200 pounds (90 kg). They first enter the water with their mothers when they are one week old, so they are strong swimmers by the time they are weaned, and are familiar by then with the routine of using breathing holes and keeping them open.

Australian Sea Lion (*Neophoca cinerea*)

This sea lion lives in the coastal waters of western and southern Australia only. It had a much wider range, but the eastern populations were exterminated by seal hunters in the nineteenth century. There are believed to be only about 12,000 animals left and they are in decline, due mainly to high pup mortality, resulting from disturbance at their breeding beaches, the development of fish farms, and their entanglement in fishing nets. They are dark brown animals, males having a patch of blond fur on the head and the females a silver-gray patch. Mature bulls weigh 660 pounds (300 kg), and reach a length of 8 feet (2.5 m). Females weigh only 220 pounds (100 kg) and are 6 feet (1.8 m) long. Typical of the sea lions, they are social creatures and live in groups of up to 15 animals that breed on sandy and pebbly beaches. The bulls are polygynous and very possessive, and females that stray beyond the territorial boundaries are driven back. Unlike other sea lions, however, they have no set breeding season. Colonies vary in their calving times; some females breed alternately in winter or summer, and others may completely miss a year.

Australian sea lion pups weigh 15 pounds (7 kg) at birth and have a chocolate-brown coat, but they molt this at the age of five months, when they grow their adult coat. The mother nurses her pup for 10 days, then goes to sea for two days and returns for one day. This pattern is then followed for 18 months, even longer by those that miss a year's breeding, and females have been seen nursing a yearling pup as well as their latest baby. Pups learn to swim in the shallows and in rock pools, and then follow their mothers into the sea when they are about three months old. Mothers have also been observed allowing other pups to nurse, in what is termed "cooperative raising." But although the herd bulls protect their territories and indirectly their own calves, they also kill pups, possibly the briefly unprotected offspring of rival males, or just pups that wandered into the bull's territory. Typical of the sea lions, Australian sea lion milk is low in solids compared to that of the seals, having just 30 percent fat and 10 percent protein. This is due to the long maternal dependence of the pup, which requires lower quality milk.

California Sea Lion (*Zalophus californianus*)

The most familiar pinniped, the California sea lion is a very intelligent and adaptable animal, trained to perform in circuses, oceanaria, zoos, and for underwater military purposes. It is a highly sociable species, congregating in large numbers on beaches and rocky shorelines, marinas, and wharves throughout the northern Pacific Ocean, from British Columbia to Mexico. One subspecies lives on the Galapagos Islands, and another occurred on the coast of Japan but is believed extinct.

California Sea Lions *The familiar "pinniped" of the west coast from British Columbia to Mexico. Unlike the "ice seals," sea lion pups nurse for their first two days, then their mothers go to sea to fish for several days, then return to feed their babies, and this routine is followed for almost a year until they are weaned. Their milk is therefore not as rich as seal milk, and is the only known placental mammal milk to lack lactose.*
Photo: Rebecca Picard, Shutterstock.com

In May and June, sea lion breeding "rookeries" may contain thousands of animals, and at this time the mature bulls aggressively defend their territories, and control up to 15 females. After a gestation period of 11 months, the females give birth within a couple of days of their arrival at the breeding beach, and are mated three weeks later. Male sea lions weigh up to 850 pounds (390 kg) and are 7 feet (2.1 m) long; the much smaller females weigh only 220 pounds (110 kg) and are 6 feet (1.8 m) long. They have a streamlined body, with a layer of blubber beneath their short coat for warmth and buoyancy. Their muzzles are pointed and doglike, and their nostrils close as soon as water touches them. Sea lions are fast and agile swimmers, and their hind flippers can be turned forward under the body, allowing them to walk on land. Adults are tan or chocolate brown, and mature males develop a large bony crest on top of their heads. They eat a variety of fish, especially hake, herring, and anchovy, plus squid and octopus.

California sea lion pups are dark brown at birth but molt when four months old, and then have a pale brown or silvery-brown coat until they get their adult coat. They are usually born on land, but births do occur in the water. They are highly precocial and are soon mobile, and their movements about the beach become longer as they mature. The pups and their mothers recognize each other through their special calls and smell, but with a thousand babies on a breeding beach, they occasionally become separated; and they may then be fostered by another female. Pups nurse for their first two days, but their mothers then go to sea to feed for several days, returning for one day to nurse them, and then returning to the sea. Males migrate along the coast during the winter, but females and their pups stay near the breeding beaches until the pups are weaned, just before the next breeding season begins. Their milk lacks lactose and contains between 32 and 35 percent fat and 9 percent protein.

Antarctic Fur Seal (*Arctocephalus gazella*)

This pinniped lives in the Southern Ocean and congregates to breed on seasonally ice-free sub-Antarctic islands south of the Antarctic Convergence—the region where the cold surface water flowing north from the Antarctic meets the warmer sub-Antarctic waters flowing south, about 1,000 miles (1,610 km) north of the continent. This includes the islands of South Georgia, South Shetland and South Orkney, plus MacQuarie Island north of the convergence. Commercial sealing in the eighteenth and nineteenth centuries virtually exterminated them, leaving only a few thousand early in the last century. But with strict international protection they have recovered, and they are now believed to number about four million. They are a small species; adult males average only 396 pounds (180 kg) in weight, and measure 6 feet (1.8 m) in length, and females weigh 99 pounds (45 kg) and are about 55 inches (1.4 m) long. They have a dense dark ginger coat, with short fur fiber. They are also unusual in eating mostly krill, the shrimplike marine invertebrates that are the major component of zooplankton, and in southern waters are represented mainly by the 2½-inch-long (6.5 cm) Antarctic krill (*Euphausia superba*). Antarctic fur seals assemble in huge breeding colonies, mostly on the island of South Georgia, where the "beachmaster" males are the first to arrive and establish and defend an area of beach from the water's edge back to the vegetation. They do not feed while they are defending their section of the beach, as they cannot leave for fear of losing it. They do not gather a harem of females in typical sea lion fashion; the females select a male, are mated, and then leave the territory.

Antarctic sea lion pups weigh 12 pounds (5.4 kg) at birth, and have a black coat. Several molts result in first a silvery-gray coat, then a brownish-gray one, and then finally the dark gingery adult coat. They are suckled for four months, on milk containing 43 percent fat and about 10 percent protein; and in typical fur seal (and sea lion) fashion, the mother returns to the sea to feed about 10 days after giving birth, swimming up to 100 miles (160 km) to the best feeding grounds. She then returns to feed her pup, and until it is weaned she follows this routine of alternately nursing her pup and fishing for krill.

Walrus (*Odobenus rosmarus*)

Walruses are the most easily identified pinnipeds, as they are the only ones with large tusks, which are present in both sexes. There are two subspecies, the Pacific walrus and the Atlantic walrus. The Pacific walrus lives in the northern waters of the Bering, Chukchi, and Laptev seas. It is the largest walrus, and males may weigh 3,748 pounds (1,700 kg) and reach a length of 12 feet (3.6 m). Females weigh 2,756 pounds (1,250 kg) and are 10 feet (3.1 m) long. The smaller Atlantic walrus occurs in northeastern Canada and Greenland, and mature bulls weigh 2,000 pounds (908 kg) and are 9.5 feet (2.9 m) long. Females weigh 1,750 pounds (795 kg) and grow up to 8 feet (2.4 m) long. Only the sea elephants are larger than these animals.

Walruses have a fusiform, or sausage-shaped, body and a cinnamon-brown coat; but blood vessels under the skin dilate during warm weather and give the animal a pink appearance. They have short and square fore-flippers and triangular-shaped hind flippers, which are used for propulsion in the water. They can rotate their hind flippers beneath their pelvic girdle (like the sea lions) and can then walk of all fours. The male's tusks may be 39 inches (1 m) long and the female's 31 inches (80 cm). They are used for social dominance and also to haul out onto the ice.

Walruses have small eyes and lack external ear flaps, and their very tough hide may be 1.6 inches (4 cm) thick. They prefer shallow water, no deeper than 260 feet (80 m), with a gravel bottom; there they search for the mollusks, especially clams, that form the bulk of their diet, locating them with the numerous stiff vibrissae or whiskers on their snouts. They either suck the clams out of their shells or crush the shells with their molars. They also eat octopus, squid, and crustaceans, as well as marine worms, which they uncover from the sand and pebbles by blowing a jet of water from their mouths.

The walrus is gregarious and lives in huge herds on the coastal shores and pack ice. It has an unusual polygynous breeding arrangement, similar to the lek system practiced by several birds. Females form large groups on the ice, and the males swim past and display by making clucking and whistling sounds underwater. When a female selects a male she joins him, is mated in the water, then returns to the ice while the male attempts to lure another female. The single calf weighs 132 pounds (60 kg), and is born on the ice but can swim at birth, so it may be nursed in the water, and it sometimes rides on its mother's back. At birth it has a short and dense soft coat of pale gray to brown, having shed its white lanugo or fetal coat about two months before it was born. There is a very strong bond between mother and calf, and lactating females also adopt orphans. Pups may begin to eat solids at six months but are dependent on milk for two years, and with their mothers they form nursery herds, separate from the bulls and other cows.

Glossary

Aberrant
Animals that deviate in important characteristics from their nearest relatives.

Adaptive radiation
The diversification of a species as it adapts to different ecological niches and eventually becomes so specialized for the new environment that it evolves into new species.

Alloparenting
The altruisitic behavior of "helpers" who assist bird and mammal parents to raise their young and so improve reproductive success. They are usually other members of the family group, such as older siblings and nonbreeding females.

Altricial
Young animals that are poorly developed at birth or hatching. They are usually naked, blind, and immobile and cannot thermoregulate, so are therefore confined to a nest, den, or the marsupial's pouch.

Amnion
The membranous sac filled with transparent fluid that encloses and protects the embryo or fetus.

Amniote
A member of the *Amniota*—vertebrates (reptiles, birds, and mammals) that have an amnion during their embryonic development.

Aposematic
Coloration or other features that warn of noxious properties, such as the poison glands of some salamanders and the poison arrow frogs.

Axil
The angle formed between a leaf and the plant stem. The water that collects there acts as a nursery pool for tropical tree frog larvae.

Barbel
Tactile fingerlike projections on the mouths of certain fish, such as the catfish, that are used to locate food.

Batrachotoxins
Very potent heart and nerve toxins produced by the poison arrow frogs.

Brood patch
A bare patch that allows birds to have direct contact with their eggs without feathers preventing heat transmission. Most female birds (and some males that also incubate the eggs) develop a brood patch during the breeding season. Changes in their hormone levels result in the down feathers on the bird's stomach falling out, or loosening and being pulled out. Ducks actually use them to line their nests. Expansion of the blood vessels to the bare patch then increases the transference of heat to the eggs and incubation can proceed.

Brooding
When birds sit on their nestlings to keep them warm. It is being increasingly used, however, to describe the hatching of the eggs, which is really "incubation." It also describes fish that hold their eggs in their mouths or in stomach pouches, although they are only providing protection, not warmth.

Carapace
The top shell of a turtle or tortoise, which may be hard and horny keratin, or soft and leathery.

Cloaca
The organ into which an animal's digestive, urinary, and reproductive systems empty, opening via the anus.

Clutch
A series of eggs laid by a reptile or bird and incubated at one time.

Convergent evolution
The development of similar structures in distantly related animals as a result of adapting to similar environments or life strategies. The evolution of the marsupials in Australia, for example, has produced many species that resemble the placental mammals in form and behavior.

Crèche
A group of young animals, all of similar age, that stay together for their protection, usually under the supervision of a few adults. This allows their parents to concentrate on such activities as gathering food for their chicks.

Cryoprotectant
Antifreeze, produced by some northern animals to prevent the freezing of their intracellular fluids.

Delayed implantation
A delay in the implantation of the fertilized egg in the wall of the uterus. It occurs in the bears, weasels, and pinnipeds, allowing the birth of their young at a more favorable time.

Ectotherm
A cold-blooded or poikilothermic animal, dependant on the environment for its body heat.

Embryo
An animal in the early stage of development before birth or hatching from an egg.

Endotherm
A warm-blooded (homeothermic) animal that can maintain its own body temperature irrespective of that of the environment.

Estivation
Long-term summer sleep to avoid hot and dry weather and to conserve water and energy, while surviving upon body fat.

Estrogen
The primary female sex hormone produced by the ovaries, and responsible for the development of secondary sexual characteristics.

Eusocial
Cooperative care involving a dominant "queen" responsible for all the breeding, and a sterile working class. It is more common in invertebrates, the only eusocial mammal being the naked mole rat.

Eutheria
The subclass of placental mammals—all species above the monotremes and the marsupials.

Facultative
Organisms that survive in either the presence or absence of a specific factor, so they are not strictly dependent upon it. Facultative parasitic birds, like the black-billed cuckoo, lay eggs in other bird's nests, but also nest and raise their own young.

Farrowing
The process of a sow giving birth to a litter of piglets.

Fratricide—see Siblicide

Gametes
Mature male or female reproductive cells—sperm or eggs.

Hatching synchronization
Some birds, such as the ostrich and rhea, lay their eggs over many days, but the chicks must leave the nest soon after they hatch for their safety and to feed. The embryonic chicks therefore call to each other in the egg and synchronize their hatching, shortening the incubation period of some eggs by several days.

Heterotherm
An animal that can maintain its temperature at a certain level when active, but at other times it fluctuates with the environment.

Hibernate
Long-term torpidity to escape unfavorable winter conditions and to conserve energy, while surviving upon body fat or external food stores.

Immunoglobulins
Proteins, also known as antibodies and gamma globulins, that are released into the bloodstream in response to infections and are present in the colostrum, or first milk, of mammals.

Implacentals
Mammals lacking a placenta—as in the monotremes and marsupials.

Imprinting

A form of early learning in birds and mammals during a critical period of their lives. It may result in an irreversible attachment to a foster parent, even a human one.

Incubation

The process of warming eggs to develop their embryos. Occurring in birds, via their brood patch, their feet, or in soil heated by decomposition, radiant, or geothermal heat. Also in the monotremes—in the echidna's pouch or when the platypus curls around her eggs—and in some snakes that can raise their temperature through muscle contractions.

Infanticide

The killing of young animals by adults. These may be their offspring or those of others.

Jacobsen's organ

An extrasensory organ in the roof of the mouth in many animals; also known as the vomeronasal organ. It receives scent particles carried by the air or on the tongue and transmits the information to the brain for action.

Lactogenesis

The production of milk by the mammary glands.

Lanugo

A fine, downy coat of hair that covers the pinniped fetus and the newborn pup.

Lateral line

A sense organ in fish and some amphibians, visible in fish as faint lines along the sides of their bodies from gills to tail, that detects movement and changes in the chemical composition of the surrounding water. The sense receptors within the line are called neuromasts.

Lek

A courtship ground where male birds and mammals of several species display to attract females.

Litter

The offspring at one birth of a mammal. Generally used only for species that have several young.

Mammary glands

The glands present in female mammals that produce milk for suckling the young. They evolved from sweat glands.

Marsupium

The pouch in which the prematurely born and helpless young of the kangaroos, possums, and related species of marsupials are raised.

Maternal

Relating to or derived from one's mother.

Metabolic rate

The rate at which an organism transforms food into energy and body tissue. The amount of energy liberated per unit of time.

Metabolism

The chemical processes within the body that sustain life. Some substances are broken down from ingested foods to provide energy, while others are synthesized internally by the animal. Two main processes are involved—anabolism, which is the biosynthesis of complex organic substances from simpler ones; and catabolism, the breakdown of complex substances to release their energy.

Metamorphosis
The process of changing from one form to another, such as the transformation of a tadpole into a froglet.

Metatheria
A subclass of the mammals containing the pouched mammals or marsupials.

Milt
A sperm-filled milky substance released by male fish to fertilize the eggs.

Monogamous
Animals that have only one mate at a time.

Mouth brooder
Fish that hold their eggs, and then the fry or hatchlings, in their mouths for their protection. They do not provide warmth.

Neoteny
The retention of larval characters into adulthood. The amphibian larval form (the tadpole in frogs and toads) is ideally adapted for an aquatic existence, and species that elected to stay in water as adults therefore retained this form. The axolotl is the classic example of this phenomenon, retaining its gills and long tail as an adult. It is also known as paedomorphism.

Nidicolous birds
Young birds that remain in the nest for some time after hatching, but need not be naked (psilopaedic) or helpless (altricial).

Nidifugous birds
Young birds that leave the nest soon after hatching. They therefore have varying degrees of independence (precocial) and have feathers or down (ptilopaedic).

Obligate
An animal that cannot survive without the assistance of others. Obligate parasitic birds, such as the European cuckoo, rely totally on other birds to hatch their eggs and raise the chicks.

Osmosis
The diffusion of fluids from a dilute solution to a concentrated one through a semipermeable membrane.

Oviparity
Reproduction in which the young hatch from eggs that have been laid, and are then incubated by the parent or foster parent, or by the envionment.

Ovoviviparity
Reproduction in which the development of the embryo occurs within the mother, and the eggs hatch just before or during the act of being laid, so it is considered live-birth. The egg yolk nourishes embryonic growth without assistance from the mother (*See* Oviparity and Viviparity).

Paedomorphism—see Neoteny

Parasitic bird
A bird that lays its eggs in other bird's nests, usually one per nest, and has no further involvement in the reproductive process. The host bird hatches the egg and raises the chick.

Parotoid glands
Large skin glands situated on the neck, sides of the head, or shoulders of many species of toads, which are reservoirs for a highly toxic milky substance.

Parthenogenesis
Development of an egg without fertilization—"virgin birth." It occurs in some species of lizards, salamanders, and fish.

Palatal valve
A valve in the palate that allows crocodiles to breathe while holding prey and while they are floating with just their nostrils above the surface.

Paternal
Relating to the father.

Pheromones
Chemical signals that travel between organisms as a form of communication to trigger a social or sexual response.

Physiology
The branch of biology dealing with the functioning of organisms.

Phytotelmata
The tiny pools of water that collect in tree holes or the axils of treetop plants, especially bromeliads, that act as nurseries for many species of small arboreal frogs.

Pituitary
An endocrine gland situated at the base of the brain that secretes important hormones, including the growth hormone.

Placental
Mammals that nourish their young in the womb through a placenta—an organ comprising embryonic and maternal tissues in union.

Plastron
The lower shell of a turtle or tortoise, hinged in some species and thus allowing them to close up tightly, after withdrawing their head, limbs, and tail.

Poikilotherm
A cold-blooded animal whose body temperature varies in accordance with its surroundings, therefore all the vertebrates except birds and mammals.

Polyandry
The form of polygamy in which a female mates with several males.

Polygamy
Animals that have multiple mates of the opposite sex.

Polygynandry
Promiscuity, when several males and several females mate indiscriminately.

Polygyny
The form of polygamy in which a male mates with several females.

Precocial
Young animals that are well developed, mobile, have their eyes open, and have a body covering of down feathers or fur, at birth or hatching. They may be independent, semi-independent or totally dependent upon their parents.

Progesterone
A female hormone produced by the ovaries after ovulation, which prepares the lining of the uterus for implantation of the fertilized egg.

Prolactin
A hormone secreted by the pituitary gland that stimulates breast development and milk production.

Promiscuous—see Polygynandry

Prototheria
The monotremes (the echidnas and the duck-billed platypus) that are the most primitive of the mammals and still retain several reptilian characteristics, including laying eggs.

Psilopaedic
Young birds that are naked or have sparse down when they hatch.

Ptilopaedic
Young birds that are covered with down when they hatch.

Radiation—see Adaptive radiation

Scutes
Large scales of horny keratin on the outer layer of turtles' shells.

Siblicide
The killing of a sibling. It occurs in several birds, especially raptors and skuas, but is only known in one mammal—the spotted hyena.

Spermatophore
A small packet of sperm produced by some animals, such as salamanders, that have internal fertilization.

Spermatogenesis
The process of the formation of spermatozoa (sperm).

Taxonomy
The scientific naming and classification of living organisms (plants and animals), begun by Swedish biologist Carl Linnaeus (1707–1778).

Tetrapod
A vertebrate with four limbs.

Thermogenesis
The production of heat in warm-blooded animals by increasing the metabolic rate and the breaking down of fat molecules.

Thermoregulation
The maintenance or regulation of normal body temperature in mammals and birds.

Trophic
Pertaining to food or nutrition. The eggs that some frogs lay purely for feeding to their tadpoles are called trophic eggs.

Viviparity
Producing eggs that are fertilized and develop within the mother's body, being nourished by the egg yolk and secretions from the uterus via a placental structure. The young are then born alive.

Vomeronasal organ—see Jacobsen's organ

Yolk sac
A compartment in the amniotic egg that contains stored food for the developing embryo.

Bibliography

Alvarez del Toro, M. On the biology of the American Finfoot in southern Mexico. *The Living Bird* 10 (1971): 79–88.

Animal Diversity Web. http://animaldiversity.ummz.umich.edu/site.

Animal Facts. http://www.bbc.co.uk/nature/wildfacts/.

Anon. *The UFAW handbook on the care and management of laboratory animals.* Livingstone, Edinburgh & London, 1966.

Ask the Experts: Biology. http://www.sciam.com/askexpert_question.

BBC Wildfacts. http://www.bbc.co.uk/nature/wildfacts/.

Bickford, D. The ecology and evolution of parental care in the Microhylid frogs of New Guinea. *Nature* 418 (2002): 601–602.

Biology Dictionary. http://www.biology-online.org/dictionary.

Bolan, E.K. Reproductive Guilds of Fishes: a proposal and definition. *Jour. Fish. Res. Board of Canada* 32 (1975): 821–864.

Breeder, C.M., Jr., and Rosen, D.E. *Modes of reproduction in fishes.* The Natural History Press, New York, 1966.

Caecilian diversity. https:///www.fiu.edu/~acaten01/caediv.html.

Calhoun, J.B. Population density and social pathology. *Scientific American* 206 (1962): 139–148.

Campbell, B., and Lack, E. (Eds.) *A Dictionary of Birds.* Buteo, Vermilion, 1985.

Clutton-Brock, T.H. *The Evolution of Parental Care.* Princeton University Press, Princeton, New Jersey, 1991.

Collins, L.R. *Monotremes and Marsupials. A reference for zoological institutions.* Smithsonian Inst. Press. Washington. 1973.

Crocodiles. AG Krokodile. The Crocodile Working Group. http://members.aol.com.agkrokodile/e_index.html.

Crocodilians. http://www.flmnh.ufl.edu/cnhc/cir.html.

Dobie, J. Frank. *The Longhorns.* Castle Books, New Jersey, 1961.

Duellman, W.E. Reproductive Strategies in Frogs. *Scientific American* 267 (1992): 58–65.

Ewer, R.F. *Ethology of mammals.* Plenum, New York. 1968.

Gibson, Richard C., and Buley, Kevin R. Maternal Care and Obligatory Oophagy in Leptodactylus fallax: A New Reproductive Mode in Frogs. *Copeia* 4 (1), (2004): 128–135.

Gemmell, R. T., Veitch, C., and Nelson, J. Birth in Marsupials. *Comp. Biochem. and Phys. B.* 131 (2002): 621–630

Gould, E., and Eisenberg, J. Notes on the biology of the Tenrecidae. *J. Mamm.* 47 (1966): 660–686.

Griffiths, M. E. *Echidnas.* Pergamon Press, Oxford. 1968.

Gubernick, D. J., and Klopfer, P. H. (Eds.) *Parental Care in Mammals.* Plenum. NY. 1981.

Hayssen, V. Empirical and Theoretical constraints on the evolution of Lactation. *Jour. Dairy Science* 76, no. 10 (1993): 3213–3233.

Hurley, Walter. T. *Milk Composition and Synthesis Resource Library.* 2006. http://classes.uiuc/edu/AnSci308.

Hrdy, S. *The langurs of Abu.* Harvard University Press, Cambridge, 1977.

Jameson, D. L. Life History and Phylogeny. *Systematic Zoology* 6 (1957): 75–80.

Jensen, R. G. (Ed.) *Handbook of Milk Composition.* Academic Press, San Diego, 1995.

Jonna, R. Cichlidae. Animal Diversity Web. 2004. http://animaldiversity.ummz.umich.edu/site/accounts/information/Cichlidae.html.

King, J. E. *Seals of the World.* British Museum of Natural History, London, 1983.

Kleiman, D. G. Monogamy in mammals. *Q. Rev. Biol.* 52 (1) (1977): 30–69.

Kleiman, D. G., Allen, M. E., Thompson, K. V., and Lumpkin, S. (Eds.) *Wild Mammals in Captivity.* University of Chicago Press, Chicago, 1996.

Lehtinen, Richard M. Parental care and reproduction in two species of mantidactylus (Anura: Mantellidae). *Journal of Herpetology* 37 (4) (2003): 766–768.

Marsupial Milk Replacers. http://www.wombaroo.com.au.

Merck Veterinary Manual. http://www.merckvetmanual.com.

Moore, J. Population density, social pathology and behavioural ecology. *Primates* 40 (1999): 5–26.

McCrane, M. P. Birth, behaviour and development of a hand-reared two-toed sloth *Choloepus didactylus. Int. Zoo Yrbk.* VI (1966): 153–163.

Nowak, R. M. *Walker's Mammals of the World.* Vols. I & II. Johns Hopkins University Press, Baltimore and London, 1991.

Pilson, M. E. Q., and Kelly, A. L. Composition of the milk from *Zalophus californianus* the California sea lion. *Science* 135 (1962):104–105.

Primates. http://www.Animalomnibus.com/primates.htm.

Primate Info Net. http://pin.primate.wisc.edu/factsheets.

Pusey, Anne E., and Packer, Craig. Non-offspring nursing in social carnivores: minimizing the costs. *Behav. Ecol.* 5 (4) (1994): 362–374.

Renouf, D. (Ed.) *Behaviour of Pinnipeds.* Chapman and Hall, London, 1991.

Reynolds, John D., Goodwin, Nicholas B., and Freckleton, Robert P. Evolutionary transitions in parental care and live bearing in vertebrates. *Philosophical Transactions of the Royal Society.* London, 2002. http://evolve.zoo.ox.ac.uk/papers/Trans2002.pdf.

Skutch, A. *Parent Birds and Their Young.* University of Texas Press, Austin, 1976.

Species information and classification. ADW: http://animaldiversity.ummz.umich.edu/site.

Steel, R. *Crocodiles.* Helm, London, 1989.

Stonehouse, B., and Gilmore, D. (Eds.) *The Biology of Marsupials.* University Park Press, Baltimore, 1977.

Tattersal, I. *The Primates of Madagascar.* Columbia University Press, New York, 1982.

Tilden, C. D., and Oftedal, O. T. Milk composition reflects patterns of maternal care in primates. *American Journal of Primatology* 41 (3) (1997): 195–211.

Trewevas, E. Tilapiine Fishes Of The Genera Sarotherodon, Oreochromis And Danakilia. *British Museum Of Natural History.* Publ. Num. 878. Comstock, Ithaca, 1983.

Widdowson, E. M. Milk and the newborn animal. *Proc. Nutr. Soc.* 43 (1984): 87–100.

Wild Carnivores. *Lioncrusher's Domain.* http://www.lioncrusher.com.

Woolley, P. Reproduction in the Antechinus spp and other dasyurid marsupials. *Symp. Zoo. Soc. Lond.* 15 (1966): 281–294.

Index

About the Author

CLIVE ROOTS has been a zoo director for many years. He has traveled the world collecting live animals for zoo conservation programs. Roots has acted as a master-planning and design consultant for numerous zoological gardens and related projects around the world, and has written many books on zoo and natural history subjects.